THE RED BOOK
OF MATHEMATICAL PROBLEMS

KENNETH S. WILLIAMS
KENNETH HARDY

Carleton University, Ottawa

Dover Publications, Inc.
Mineola, New York

Bibliographical Note

This Dover edition, first published in 1996, is a slightly corrected republication of the work originally published by Integer Press, Ottawa, Canada, in 1988 under the title *The Red Book: 100 Practice Problems for Undergraduate Mathematics Competitions*. A section of the original page 97 has been deleted and all subsequent copy repaged thereafter.

Library of Congress Cataloging-in-Publication Data

Williams, Kenneth S.
    The red book of mathematical problems / Kenneth S. Williams, Kenneth Hardy.
        p.   cm.
    "A slightly corrected republication of the work originally published by Integer Press, Ottawa, Canada, in 1988 under the title: The red book: 100 practice problems for undergraduate mathematics competitions"—T.p. verso.
    Includes bibliographical references.
    ISBN 0-486-69415-1 (pbk.)
    1. Mathematics—Problems, exercises, etc.   I. Hardy, Kenneth.
II. Title.
QA43.W55   1996
510′.76—dc20   96-43820
                                                                      CIP

Manufactured in the United States of America
Dover Publications, Inc., 31 East 2nd Street, Mineola, N.Y. 11501

PREFACE TO THE FIRST EDITION

It has become the fashion for some authors to include literary quotations in their mathematical texts, presumably with the aim of connecting mathematics and the humanities. The preface of *The Green Book** of 100 practice problems for undergraduate mathematics competitions hinted at connections between problem-solving and all the traditional elements of a fairy tale: mystery, search, discovery, and finally resolution. Although *The Red Book* may seem to have political overtones, rest assured, dear reader, that the quotations (labelled Marx, Pushkin and Trotsky, just for fun) are merely an inspiration for your journey through the enchanted realms of mathematics.

The Red Book contains 100 problems for undergraduate students training for mathematics competitions, particularly the William Lowell Putnam Mathematical Competition. Along with the problems come useful hints, and complete solutions. The book will also be useful to anyone interested in the posing and solving of mathematical problems at the undergraduate level.

Many of the problems were suggested by ideas originating in a variety of sources, including *Crux Mathematicorum, Mathematics Magazine* and the *American Mathematical Monthly*, as well as various mathematics competitions. Where possible, acknowledgement to known sources is given at the end of the book.

Once again, we would be interested in your reaction to *The Red Book,* and invite comments, alternate solutions, and even corrections. We make no claim that the solutions are the "best possible" solutions, but we trust that you will find them elegant enough, and that *The Red Book* will be a practical tool in training undergraduate competitors.

We wish to thank our typesetter and our literary adviser at Integer Press for their valuable assistance in this project.

Kenneth S. Williams and Kenneth Hardy

Ottawa, Canada

May, 1988

*To be reprinted by Dover Publications in 1997.

To our children

Helen, Jonathan, and Joanna
— K.S.W.

Shauna Eliot and Angelica Kate
— K.H.

CONTENTS

Page

NOTATION

$[x]$	denotes the greatest integer $\leq x$, where x is a real number.
$\ln x$	denotes the natural logarithm of x.
$\exp x$	denotes the exponential function e^x.
$\phi(n)$	denotes Euler's totient function defined for any natural number n.
$GCD(a, b)$	denotes the greatest common divisor of the integers a and b.
$\binom{n}{k}$	denotes the binomial coefficient $n!/k!\,(n-k)!$, where n and k are non-negative integers (the symbol having value zero when $n < k$).
$\left(\dfrac{n}{p}\right)$	denotes Legendre's symbol which has value $+1$ (resp. -1) if the integer n is a quadratic residue (resp. nonresidue) modulo the odd prime p.
$\deg\,(f(x))$	denotes the degree of the polynomial $f(x)$.
$\underline{v}^t$	denotes the transpose of the row vector $\underline{v}$.
$\tau(n)$	denotes the number of distinct prime divisors of the positive integer n.
$f'(x)$	denotes the derivative of the function $f(x)$ with respect to x.
$\det A$	denotes the determinant of the square matrix A.
Z	denotes the domain of rational integers.
Q, R, C	denote the fields of rational, real, complex numbers respectively.

THE PROBLEMS

Mankind always sets itself only such problems as it can solve; ... it will always be found that the task itself arises only when the material conditions for its solution already exist or are at least in the process of formation.

Karl Marx (1818-1883)

1. Let p denote an odd prime and set $\omega = \exp(2\pi i/p)$. Evaluate the product

$$(1.0) \qquad E(p) = (\omega^{r_1} + \omega^{r_2} + \ldots + \omega^{r_{(p-1)/2}})(\omega^{n_1} + \omega^{n_2} + \ldots + \omega^{n_{(p-1)/2}}),$$

where $r_1, \ldots, r_{(p-1)/2}$ denote the $(p-1)/2$ quadratic residues modulo p and $n_1, \ldots, n_{(p-1)/2}$ denote the $(p-1)/2$ quadratic nonresidues modulo p.

2. Let k denote a positive integer. Determine the number $N(k)$ of triples (x, y, z) of integers satisfying

$$(2.0) \qquad \begin{cases} |x| \leq k, & |y| \leq k, & |z| \leq k, \\ |x - y| \leq k, & |y - z| \leq k, & |z - x| \leq k. \end{cases}$$

3. Let $p \equiv 1 \pmod 4$ be prime. It is known that there exists a unique integer $w \equiv w(p)$ such that

$$w^2 \equiv -1 \pmod p, \quad 0 < w < p/2.$$

(For example, $w(5) = 2, w(13) = 5$.) Prove that there exist integers a, b, c, d with $ad - bc = 1$ such that

$$pX^2 + 2wXY + \frac{(w^2 + 1)}{p}Y^2 \equiv (aX + bY)^2 + (cX + dY)^2.$$

(For example, when $p = 5$ we have

$$5X^2 + 4XY + Y^2 \equiv X^2 + (2X + Y)^2,$$

and when $p = 13$ we have

$$13X^2 + 10XY + 2Y^2 \equiv (3X + Y)^2 + (2X + Y)^2.)$$

4.   Let $d_r(n)$, $r = 0, 1, 2, 3$, denote the number of positive integral divisors of n which are of the form $4k + r$. Let m denote a positive integer. Prove that

(4.0) $$\sum_{n=1}^{m}(d_1(n) - d_3(n)) = \sum_{j=0}^{\infty}(-1)^j \left[\frac{m}{2j + 1}\right].$$

5.   Prove that the equation

(5.0) $$y^2 = x^3 + 23$$

has no solutions in integers x and y.

6.   Let $f(x, y) = ax^2 + 2bxy + cy^2$ be a positive-definite quadratic form. Prove that

(6.0) $$(f(x_1, y_1)f(x_2, y_2))^{1/2}f(x_1 - x_2, y_1 - y_2)$$
$$\geq (ac - b^2)(x_1y_2 - x_2y_1)^2,$$

for all real numbers x_1, x_2, y_1, y_2.

7.   Let R, S, T be three real numbers, not all the same. Give a condition which is satisfied by one and only one of the three triples

(7.0)
$$\begin{cases} (R, S, T), \\ (T, -S + 2T, R - S + T), \\ (R - S + T, 2R - S, R). \end{cases}$$

8.   Let $ax^2 + bxy + cy^2$ and $Ax^2 + Bxy + Cy^2$ be two positive-definite quadratic forms, which are not proportional. Prove that the form

(8.0) $(aB - bA)x^2 + 2(aC - cA)xy + (bC - cB)y^2$

is indefinite.

9.   Evaluate the limit

(9.0)
$$L = \lim_{n \to \infty} \frac{n}{2^n} \sum_{k=1}^{n} \frac{2^k}{k} .$$

10.   Prove that there does not exist a constant $c \geq 1$ such that

(10.0) $n^c \phi(n) \geq m^c \phi(m),$

for all positive integers n and m satisfying $n \geq m$.

11.   Let D be a squarefree integer greater than 1 for which there exist positive integers A_1, A_2, B_1, B_2 such that

(11.0)
$$\begin{cases} D = A_1^2 + B_1^2 = A_2^2 + B_2^2, \\ \quad (A_1, B_1) \neq (A_2, B_2). \end{cases}$$

Prove that neither

$$2D(D + A_1 A_2 + B_1 B_2)$$

nor

$$2D(D + A_1 A_2 - B_1 B_2)$$

is the square of an integer.

12.   Let $\mathbf{Q}$ and $\mathbf{R}$ denote the fields of rational and real numbers respectively. Let $\mathbf{K}$ and $\mathbf{L}$ be the smallest subfields of $\mathbf{R}$ which contain both $\mathbf{Q}$ and the real numbers

$$\sqrt{1985 + 31\sqrt{1985}} \quad \text{and} \quad \sqrt{3970 + 64\sqrt{1985}} \,,$$

respectively. Prove that $\mathbf{K} = \mathbf{L}$.

13.   Let k and l be positive integers such that

$$GCD(k, 5) = GCD(l, 5) = GCD(k, l) = 1$$

and

$$-k^2 + 3kl - l^2 = F^2, \quad \text{where } GCD(F, 5) = 1.$$

Prove that the pair of equations

(13.0) $\begin{cases} k = x^2 + y^2, \\ l = x^2 + 2xy + 2y^2, \end{cases}$

has exactly two solutions in integers x and y.

14.   Let r and s be non-zero integers. Prove that the equation

(14.0) $(r^2 - s^2)x^2 - 4rsxy - (r^2 - s^2)y^2 = 1$

has no solutions in integers x and y.

15. Evaluate the integral

(15.0)
$$I = \int_0^1 \ln x \ln(1 - x) \, dx .$$

16. Solve the recurrence relation

(16.0)
$$\sum_{k=1}^{n} \binom{n}{k} a(k) = \frac{n}{n+1} , \quad n = 1, 2, \ldots .$$

17. Let n and k be positive integers. Let p be a prime such that

$$p > (n^2 + n + k)^2 + k .$$

Prove that the sequence

(17.0)
$$n^2, \; n^2 + 1, \; n^2 + 2, \ldots, n^2 + l ,$$

where $l = (n^2 + n + k)^2 - n^2 + k$, contains a pair of integers $(m, m + k)$ such that

$$\left(\frac{m}{p}\right) = \left(\frac{m+k}{p}\right) = 1 .$$

18. Let

$$a_n = \frac{1}{4n+1} + \frac{1}{4n+3} - \frac{1}{2n+2} , \quad n = 0, 1, \ldots .$$

Does the infinite series $\sum_{n=0}^{\infty} a_n$ converge, and if so, what is its sum?

19. Let $a_1, \ldots, a_m$ be $m \; (\geq 2)$ real numbers. Set

$$A_n = a_1 + a_2 + \ldots + a_n, \quad n = 1, 2, \ldots, m .$$

Prove that

(19.0) $$\sum_{n=2}^{m} \left(\frac{A_n}{n}\right)^2 \leq 12 \sum_{n=1}^{m} a_n^2 \ .$$

20.   Evaluate the sum

$$S = \sum_{k=0}^{n} \frac{\binom{n}{k}}{\binom{2n-1}{k}}$$

for all positive integers n.

21.   Let a and b be coprime positive integers. For k a positive integer, let $N(k)$ denote the number of integral solutions to the equation

(21.0) $ax + by = k, \quad x \geq 0, \quad y \geq 0 \ .$

Evaluate the limit

$$L = \lim_{k \to +\infty} \frac{N(k)}{k} \ .$$

22.   Let a, d and r be positive integers. For $k = 0, 1, \ldots$ set

(22.0) $u_k = u_k(a, d, r) = \dfrac{1}{(a + kd)(a + (k+1)d) \ldots (a + (k+r)d)} \ .$

Evaluate the sum

$$S = \sum_{k=0}^{n} u_k \ ,$$

where n is a positive integer.

23.   Let $x_1, \ldots, x_n$ be $n \ (> 1)$ real numbers. Set

$$x_{ij} = x_i - x_j \quad (1 \leq i < j \leq n).$$

Let F be a real-valued function of the $n(n-1)/2$ variables x_{ij} such that the inequality

(23.0) $$F(x_{11}, x_{12}, \ldots, x_{n-1\,n}) \le \sum_{k=1}^{n} x_k^2$$

holds for all $x_1, \ldots, x_n$.

Prove that equality cannot hold in (23.0) if $\sum_{k=1}^{n} x_k \ne 0$.

24. Let $a_1, \ldots, a_m$ be m (≥ 1) real numbers which are such that $\sum_{n=1}^{m} a_n \ne 0$. Prove the inequality

(24.0) $$\left(\sum_{n=1}^{m} n a_n^2 \right) \Big/ \left(\sum_{n=1}^{m} a_n \right)^2 > \frac{1}{2\sqrt{m}}.$$

25. Prove that there exist infinitely many positive integers which are not expressible in the form $n^2 + p$, where n is a positive integer and p is a prime.

26. Evaluate the infinite series

$$S = \sum_{n=1}^{\infty} \arctan\left(\frac{2}{n^2}\right).$$

27. Let $p_1, \ldots, p_n$ denote n (≥ 1) distinct integers and let $f_n(x)$ be the polynomial of degree n given by

$$f_n(x) = (x - p_1)(x - p_2)\ldots(x - p_n).$$

Prove that the polynomial

$$g_n(x) = (f_n(x))^2 + 1$$

cannot be expressed as the product of two non-constant polynomials with integral coefficients.

28.   Two people, A and B, play a game in which the probability that A wins is p, the probability that B wins is q, and the probability of a draw is r. At the beginning, A has m dollars and B has n dollars. At the end of each game the winner takes a dollar from the loser. If A and B agree to play until one of them loses all his/her money, what is the probabilty of A winning all the money?

29.   Let $f(x)$ be a monic polynomial of degree $n \geq 1$ with complex coefficients. Let $x_1, \ldots, x_n$ denote the n complex roots of $f(x)$. The discriminant $D(f)$ of the polynomial $f(x)$ is the complex number

$$(29.0) \qquad\qquad D(f) = \prod_{1 \leq i < j \leq n} (x_i - x_j)^2.$$

Express the discriminant of $f(x^2)$ in terms of $D(f)$.

30.   Prove that for each positive integer n there exists a circle in the xy-plane which contains exactly n lattice points.

31.   Let n be a given non-negative integer. Determine the number $S(n)$ of solutions of the equation

$$(31.0) \qquad\qquad x + 2y + 2z = n$$

in non-negative integers x, y, z.

32.   Let n be a fixed integer ≥ 2. Determine all functions $f(x)$, which are bounded for $0 < x < a$, and which satisfy the functional equation

$$(32.0) \qquad f(x) = \frac{1}{n^2} \left(f\left(\frac{x}{n}\right) + f\left(\frac{x+a}{n}\right) + \ldots + f\left(\frac{x+(n-1)a}{n}\right) \right).$$

33.   Let I denote the closed interval $[a, b]$, $a < b$. Two functions $f(x)$, $g(x)$ are said to be *completely different* on I if $f(x) \neq g(x)$ for all x in I. Let $q(x)$ and $r(x)$ be functions defined on I such that the differential equation

$$\frac{dy}{dx} = y^2 + q(x)y + r(x)$$

has three solutions $y_1(x)$, $y_2(x)$, $y_3(x)$ which are pairwise completely different on I. If $z(x)$ is a fourth solution such that the pairs of functions $z(x)$, $y_i(x)$ are completely different for $i = 1, 2, 3$, prove that there exists a constant K $(\neq 0, 1)$ such that

(33.0) $$z = \frac{y_1(Ky_2 - y_3) + (1 - K)y_2y_3}{(K - 1)y_1 + (y_2 - Ky_3)} .$$

34.   Let a_n, $n = 2, 3, \ldots$, denote the number of ways the product $b_1b_2 \ldots b_n$ can be bracketed so that only two of the b_i are multiplied together at any one time. For example, $a_2 = 1$ since b_1b_2 can only be bracketed as (b_1b_2), whereas $a_3 = 2$ as $b_1b_2b_3$ can be bracketed in two ways, namely, $(b_1(b_2b_3))$ and $((b_1b_2)b_3)$. Obtain a formula for a_n.

35.   Evaluate the limit

(35.0) $$L = \lim_{y \to 0} \frac{1}{y} \int_0^\pi \tan(y \sin x) \, dx .$$

36.   Let ϵ be a real number with $0 < \epsilon < 1$. Prove that there are infinitely many integers n for which

(36.0) $$\cos n \geq 1 - \epsilon .$$

37. Determine all the functions f, which are everywhere differentiable and satisfy

(37.0) $$f(x) + f(y) = f\left(\frac{x + y}{1 - xy}\right)$$

for all real x and y with $xy \neq 1$.

38. A point X is chosen inside or on a circle. Two perpendicular chords AC and BD of the circle are drawn through X. (In the case when X is on the circle, the degenerate case, when one chord is a diameter and the other is reduced to a point, is allowed.) Find the greatest and least values which the sum $S = |AC| + |BD|$ can take for all possible choices of the point X.

39. For $n = 1, 2, \ldots$ define the set A_n by

$$A_n = \begin{cases} \{0, 2, 4, 6, 8, \ldots\}, & \text{if } n \equiv 0 \pmod 2, \\ \{0, 3, 6, \ldots, 3(n-1)/2\}, & \text{if } n \equiv 1 \pmod 2. \end{cases}$$

Is it true that

$$\bigcup_{n=1}^{\infty}\left(\bigcap_{k=1}^{\infty} A_{n+k}\right) = \bigcap_{n=1}^{\infty}\left(\bigcup_{k=1}^{\infty} A_{n+k}\right) ?$$

40. A sequence of repeated independent trials is performed. Each trial has probability p of being successful and probability $q = 1 - p$ of failing. The trials are continued until an uninterrupted sequence of n successes is obtained. The variable X denotes the number of trials required to achieve this goal. If $p_k = \text{Prob}(X = k)$, determine the probability generating function $P(x)$ defined by

(40.0) $$P(x) = \sum_{k=0}^{\infty} p_k x^k .$$

41. A, B, C, D are four points lying on a circle such that $ABCD$ is a convex quadrilateral. Determine a formula for the radius of the circle in terms of $a = |AB|$, $b = |BC|$, $c = |CD|$ and $d = |DA|$.

42.   Let $ABCD$ be a convex quadrilateral. Let P be the point outside $ABCD$ such that $|AP| = |PB|$ and $\angle APB = 90^0$. The points Q, R, S are similarly defined. Prove that the lines PR and QS are of equal length and perpendicular.

43.   Determine polynomials $p(x, y, z, w)$ and $q(x, y, z, w)$ with real coefficients such that

$$(43.0) \qquad (xy + z + w)^2 - (x^2 - 2z)(y^2 - 2w)$$
$$\equiv (p(x, y, z, w))^2 - (x^2 - 2z)(q(x, y, z, w))^2 \ .$$

44.   Let $\mathbf{C}$ denote the field of complex numbers. Let $f : \mathbf{C} \to \mathbf{C}$ be a function satisfying

$$(44.0) \qquad \begin{cases} f(0) = 0 \ , \\ |f(z) - f(w)| = |z - w| \ , \end{cases}$$

for all z in $\mathbf{C}$ and $w = 0, 1, i$. Prove that

$$f(z) = f(1)z \quad \text{or} \quad f(1)\bar{z} \ ,$$

where $|f(1)| = 1$.

45.   If x and y are rational numbers such that

$$(45.0) \qquad \qquad \qquad \tan \pi x = y \ ,$$

prove that $x = k/4$ for some integer k not congruent to 2 (mod 4).

46.   Let P be a point inside the triangle ABC. Let AP meet BC at D, BP meet CA at E, and CP meet AB at F. prove that

(46.0)
$$\frac{|PA|}{|PD|}\frac{|PB|}{|PE|} + \frac{|PB|}{|PE|}\frac{|PC|}{|PF|} + \frac{|PC|}{|PF|}\frac{|PA|}{|PD|} \geq 12 .$$

47.   Let l and n be positive integers such that

$$1 \leq l < n, \quad GCD(l,n) = 1 .$$

Define the integer k uniquely by

$$1 \leq k < n, \quad kl \equiv -1 \pmod{n} .$$

Let M be the $k \times l$ matrix whose (i,j)-th entry is

$$(i-1)l + j .$$

Let N be the $k \times l$ matrix formed by taking the columns of M in reverse order and writing the entries as the rows of N. What is the relationship between the (i,j)-th entry of M and the (i,j)-th entry of N modulo n?

48.   Let m and n be integers such that $1 \leq m < n$. Let a_{ij}, $i = 1, 2, \ldots, m$; $j = 1, 2, \ldots, n$, be mn integers which are not all zero, and set

$$a = \max_{\substack{1 \leq i \leq m \\ 1 \leq j \leq n}} |a_{ij}| .$$

Prove that the system of equations

(48.0)
$$\begin{cases} a_{11}x_1 + a_{12}x_2 + \cdots + a_{1n}x_n & = 0 , \\ a_{21}x_1 + a_{22}x_2 + \cdots + a_{2n}x_n & = 0 , \\ \vdots \\ a_{m1}x_1 + a_{m2}x_2 + \cdots + a_{mn}x_n & = 0 , \end{cases}$$

has a solution in integers $x_1, x_2, \ldots, x_n$, not all zero, satisfying

$$|x_j| \leq \left[(2na)^{\frac{m}{n-m}}\right], \quad 1 \leq j \leq n .$$

49. Liouville proved that if

$$\int f(x)e^{g(x)}dx$$

is an elementary function, where $f(x)$ and $g(x)$ are rational functions with degree of $g(x) > 0$, then

$$\int f(x)e^{g(x)}dx = h(x)e^{g(x)} ,$$

where $h(x)$ is a rational function. Use Liouville's result to prove that

$$\int e^{-x^2} dx$$

is not an elementary function.

50. The sequence $x_0, x_1, \ldots$ is defined by the conditions

(50.0) $x_0 = 0, \quad x_1 = 1, \quad x_{n+1} = \dfrac{x_n + nx_{n-1}}{n+1}, \quad n \geq 1 .$

Determine

$$L = \lim_{n \to \infty} x_n .$$

51. Prove that the only integers $N \geq 3$ with the following property:

(51.0)   if $1 < k \leq N$ and $GCD(k, N) = 1$ then k is prime, are

$$N = 3, 4, 6, 8, 12, 18, 24, 30 .$$

52.   Find the sum of the infinite series

$$S = 1 - \frac{1}{4} + \frac{1}{6} - \frac{1}{9} + \frac{1}{11} - \frac{1}{14} + \cdots .$$

53.   Semicircles are drawn externally to the sides of a given triangle. The lengths of the common tangents to these semicircles are l, m, and n. Relate the quantity

$$\frac{lm}{n} + \frac{mn}{l} + \frac{nl}{m}$$

to the lengths of the sides of the triangle.

54.   Determine all the functions $H : \mathbf{R}^4 \to \mathbf{R}$ having the properties

(i) $H(1, 0, 0, 1) = 1,$
(ii) $H(\lambda a, b, \lambda c, d) = \lambda H(a, b, c, d),$
(iii) $H(a, b, c, d) = -H(b, a, d, c),$
(iv) $H(a + e, b, c + f, d) = H(a, b, c, d) + H(e, b, f, d),$

where $a, b, c, d, e, f, \lambda$ are real numbers.

55.   Let $z_1, \ldots, z_n$ be the complex roots of the equation

$$z^n + a_1 z^{n-1} + \ldots + a_n = 0 ,$$

where $a_1, \ldots, a_n$ are n (≥ 1) complex numbers. Set

$$A = \max_{1 \leq k \leq n} |a_k| .$$

Prove that

$$|z_j| \leq 1 + A, \quad j = 1, 2, \ldots, n.$$

56. If m and n are positive integers with m odd, determine

$$d = GCD(2^m - 1, 2^n + 1) \,.$$

57. If $f(x)$ is a polynomial of degree $2m + 1$ with integral coefficients for which there are $2m + 1$ integers $k_1, \ldots, k_{2m+1}$ such that

$$(57.0) \qquad f(k_1) = \ldots = f(k_{2m+1}) = 1 \,,$$

prove that $f(x)$ is not the product of two non-constant polynomials with integral coefficients.

58. Prove that there do not exist integers a, b, c, d (not all zero) such that

$$(58.0) \qquad a^2 + 5b^2 - 2c^2 - 2cd - 3d^2 = 0 \,.$$

59. Prove that there exist infinitely many positive integers which are not representable as sums of fewer than ten squares of odd natural numbers.

60. Evaluate the integral

$$(60.0) \qquad I(k) = \int_0^\infty \frac{\sin kx \, \cos^k x}{x} \, dx \,,$$

where k is a positive integer.

61.   Prove that

$$\frac{1}{n+1}\binom{2n}{n}$$

is an integer for $n = 1, 2, 3, \ldots$.

62.   Find the sum of the infinite series

$$S = \sum_{n=0}^{\infty} \frac{2^n}{a^{2^n} + 1},$$

where $a > 1$.

63.   Let k be an integer. Prove that the formal power series

$$\sqrt{1 + kx} = 1 + a_1 x + a_2 x^2 + \ldots$$

has integral coefficients if and only if $k \equiv 0 \pmod 4$.

64.   Let m be a positive integer. Evaluate the determinant of the $m \times m$ matrix M_m whose (i,j)-th entry is $GCD(i,j)$.

65.   Let l and m be positive integers with l odd and for which there are integers x and y with

$$\begin{cases} l & = x^2 + y^2, \\ m & = x^2 + 8xy + 17y^2. \end{cases}$$

Prove that there do not exist integers u and v with

(65.0) $$\begin{cases} l & = u^2 + v^2, \\ m & = 5u^2 + 16uv + 13v^2. \end{cases}$$

66. Let

$$a_n = 1 - \frac{1}{2} + \frac{1}{3} - \ldots + \frac{(-1)^{n-1}}{n} - \ln 2.$$

Prove that $\sum_{n=1}^{\infty} a_n$ converges and determine its sum.

67. Let $A = \{a_i \mid 0 \leq i \leq 6\}$ be a sequence of seven integers satisfying

$$0 = a_0 \leq a_1 \leq \ldots \leq a_6 \leq 6.$$

For $i = 0, 1, \ldots, 6$ let

$$N_i = \text{number of } a_j \ (0 \leq j \leq 6) \text{ such that } a_j = i.$$

Determine all sequences A such that

(67.0) $$N_i = a_{6-i}, \quad i = 0, 1, \ldots, 6.$$

68. Let G be a finite group with identity e. If G contains elements g and h such that

(68.0) $$g^5 = e, \quad ghg^{-1} = h^2,$$

determine the order of h.

69. Let a and b be positive integers such that

$$GCD(a, b) = 1, \quad a \not\equiv b \pmod{2}.$$

If the set S has the following two properties:

    (i) $a, b \in S$,
    (ii) $x, y, z \in S$ implies $x + y + z \in S$,

prove that every integer $> 2ab$ belongs to S.

70.   Prove that every integer can be expressed in the form $x^2+y^2-5z^2$, where x, y, z are integers.

71.   Evaluate the sum of the infinite series

$$\frac{\ln 2}{2} - \frac{\ln 3}{3} + \frac{\ln 4}{4} - \frac{\ln 5}{5} + \cdots \quad .$$

72.   Determine constants a, b and c such that

$$\sqrt{n} = \sum_{k=0}^{n-1} \sqrt[3]{\sqrt{ak^3 + bk^2 + ck + 1} - \sqrt{ak^3 + bk^2 + ck}} \, ,$$

for $n = 1, 2, \ldots$.

73.   Let n be a positive integer and a, b integers such that

$$GCD(a, b, n) = 1 \, .$$

Prove that there exist integers a_1, b_1 with

$$a_1 \equiv a \pmod{n}, \quad b_1 \equiv b \pmod{n}, \quad GCD(a_1, b_1) = 1 \, .$$

74.   For $n = 1, 2, \ldots$ let $s(n)$ denote the sum of the digits of 2^n. Thus, for example, as $2^8 = 256$ we have $s(8) = 2+5+6 = 13$. Determine all positive integers n such that

$$(74.0) \qquad\qquad\qquad s(n) = s(n + 1) \, .$$

75.   Evaluate the sum of the infinite series

$$S = \sum_{\substack{m,n=1 \\ GCD(m,n)=1}}^{\infty} \frac{1}{mn(m+n)}.$$

76.   A cross-country racer runs a 10-mile race in 50 minutes. Prove that somewhere along the course the racer ran 2 miles in exactly 10 minutes.

77.   Let AB be a line segment with midpoint O. Let R be a point on AB between A and O. Three semicircles are constructed on the same side of AB as follows: S_1 is the semicircle with centre O and radius $|OA| = |OB|$; S_2 is the semicircle with centre R and radius $|AR|$, meeting RB at C; S_3 is the semicircle with centre S (the midpoint of CB) and radius $|CS| = |SB|$. The common tangent to S_2 and S_3 touches S_2 at P and S_3 at Q. The perpendicular to AB through C meets S_1 at D. Prove that $PCQD$ is a rectangle.

78.   Determine the inverse of the $n \times n$ matrix

(78.0)
$$S = \begin{bmatrix} 0 & 1 & 1 & \cdots & 1 \\ 1 & 0 & 1 & \cdots & 1 \\ 1 & 1 & 0 & \cdots & 1 \\ \vdots & \vdots & \vdots & \ddots & \vdots \\ 1 & 1 & 1 & \cdots & 0 \end{bmatrix},$$

where $n \geq 2$.

79.   Evaluate the sum

(79.0)
$$S(n) = \sum_{k=0}^{n-1} (-1)^k \cos^n(k\pi/n),$$

where n is a positive integer.

80.   Determine 2×2 matrices B and C with integral entries such that

(80.0)
$$\begin{bmatrix} -1 & 1 \\ 0 & -2 \end{bmatrix} = B^3 + C^3 .$$

81.   Find two non-congruent similar triangles with sides of integral length having the lengths of two sides of one triangle equal to the lengths of two sides of the other.

82.   Let a, b, c be three real numbers with $a < b < c$. The function $f(x)$ is continuous on $[a, c]$ and differentiable on (a, c). The derivative $f'(x)$ is strictly increasing on (a, c). Prove that

(82.0)
$$(c - b)f(a) + (b - a)f(c) > (c - a)f(b) .$$

83.   The sequence $\{a_m \mid m = 1, 2, \ldots\}$ is such that $a_m > a_{m+1} > 0$, $m = 1, 2, \ldots$, and $\sum_{m=1}^{\infty} a_m$ converges. Prove that

$$\sum_{m=1}^{\infty} m(a_m - a_{m+1})$$

converges and determine its sum.

84.   The continued fraction of $\sqrt{D}$, where D is an odd nonsquare integer > 5, has a period of length one. What is the length of the period of the continued fraction of $\frac{1}{2}(1 + \sqrt{D})$?

85.   Let G be a group which has the following two properties:

(85.0) (i) G has no element of order 2,
$\quad\quad$ (ii) $(xy)^2 = (yx)^2$, for all $x, y \in G$.

Prove that G is abelian.

86.   Let $A = [a_{ij}]$ be an $n \times n$ real symmetric matrix whose entries satisfy

(86.0) $$a_{ii} = 1 , \quad \sum_{j=1}^{n} |a_{ij}| \leq 2 ,$$

for all $i = 1, 2, \ldots, n$. Prove that $0 \leq \det A \leq 1$.

87.   Let R be a finite ring containing an element r which is not a divisor of zero. Prove that R must have a multiplicative identity.

88.   Set $J_n = \{1, 2, \ldots, n\}$. For each non-empty subset S of J_n define

$$w(S) = \max_{s \in S} S - \min_{s \in S} S .$$

Determine the average of $w(S)$ over all non-empty subsets S of J_n.

89.   Prove that the number of odd binomial coefficients in each row of Pascal's triangle is a power of 2.

90.   From the $n \times n$ array

$$\begin{bmatrix} 1 & 2 & 3 & \ldots & n \\ n+1 & n+2 & n+3 & \ldots & 2n \\ 2n+1 & 2n+2 & 2n+3 & \ldots & 3n \\ \vdots & \vdots & \vdots & \ddots & \vdots \\ (n-1)n+1 & (n-1)n+2 & (n-1)n+3 & \ldots & n^2 \end{bmatrix}$$

a number x_1 is selected. The row and column containing x_1 are then deleted. From the resulting array a number x_2 is selected, and its row and column deleted as before. The selection is continued until only one number x_n remains available for selection. Determine the sum $x_1 + x_2 + \cdots + x_n$.

91.   Suppose that p X's and q O's are placed on the circumference of a circle. The number of occurrences of two adjacent X's is a and the number of occurrences of two adjacent O's is b. Determine $a - b$ in terms of p and q.

92.   In the triangular array

(92.0)

$$\begin{array}{ccccccccc}
 & & & & 1 & & & & \\
 & & & 1 & 1 & 1 & & & \\
 & & 1 & 2 & 3 & 2 & 1 & & \\
 & 1 & 3 & 6 & 7 & 6 & 3 & 1 & \\
1 & 4 & 10 & 16 & 19 & 16 & 10 & 4 & 1
\end{array}$$
$$\cdot \quad \cdot \quad \cdot$$

every entry (except the top 1) is the sum of the entry a immediately above it, and the entries b and c immediately to the left and right of a. Absence of an entry indicates zero. Prove that every row after the second row contains an entry which is even.

93.   A sequence of n real numbers $x_1, \ldots, x_n$ satisfies

(93.0)
$$\begin{cases} x_1 = 0, \\ |x_i| = |x_{i-1} + c| \quad (2 \leq i \leq n), \end{cases}$$

where c is a positive real number. Determine a lower bound for the average of $x_1, \ldots, x_n$ as a function of c only.

94.   Prove that the polynomial

(94.0)
$$f(x) = x^n + x^3 + x^2 + x + 5$$

is irreducible over $\mathbf{Z}$ for $n \geq 4$.

95. Let $a_1, \ldots, a_n$ be $n \ (\geq 4)$ distinct real numbers. Determine the general solution of the system of $n - 2$ linear equations

$$(95.0) \quad \begin{cases} x_1 + x_2 + \cdots + x_n & = 0, \\ a_1 x_1 + a_2 x_2 + \cdots + a_n x_n & = 0, \\ a_1^2 x_1 + a_2^2 x_2 + \cdots + a_n^2 x_n & = 0, \\ \vdots \\ a_1^{n-3} x_1 + a_2^{n-3} x_2 + \cdots + a_n^{n-3} x_n & = 0, \end{cases}$$

in the n unknowns $x_1, \ldots, x_n$.

96. Evaluate the sum

$$S(N) = \sum_{\substack{1 \leq m < n \leq N \\ m+n > N \\ GCD(m,n)=1}} \frac{1}{mn}, \quad N = 2, 3, \ldots .$$

97. Evaluate the limit

$$(97.0) \quad L = \lim_{n \to \infty} \frac{1}{n} \sum_{j=1}^{n} \sum_{k=1}^{n} \frac{j}{j^2 + k^2} .$$

98. Prove that

$$(98.0) \quad \tan \frac{3\pi}{11} + 4 \sin \frac{2\pi}{11} = \sqrt{11} .$$

99. For $n = 1, 2, \ldots$ let

$$c_n = 1 + \frac{1}{2} + \frac{1}{3} + \cdots + \frac{1}{n}.$$

Evaluate the sum

$$S = \sum_{n=1}^{\infty} \frac{c_n}{n(n+1)}.$$

100. For $x > 1$ determine the sum of the infinite series

$$\frac{x}{x+1} + \frac{x^2}{(x+1)(x^2+1)} + \frac{x^4}{(x+1)(x^2+1)(x^4+1)} + \cdots.$$

THE HINTS

Still shrouded in the darkest night, we look to the East with expectation: a hint of a bright new day.

Aleksander Sergeevich Pushkin (1799-1837)

1. Let

$$N(k) = \sum_{\substack{i,j=1 \\ r_i + n_j \equiv k \,(\mathrm{mod}\ p)}}^{(p-1)/2} 1 \,, \qquad k = 0, 1, \ldots, p-1 \,,$$

and prove that

$$N(k) = N(1) \,, \quad k = 1, 2, \ldots, p-1 \,.$$

Next, evaluate $N(0)$ and $N(1)$, and then deduce the value of $E(p)$ from

$$E(p) = \sum_{k=0}^{p-1} \omega^k N(k) \,.$$

2. Prove that

$$N(k) = \sum_x \sum_y \sum_z 1 \,,$$

where the variable x is summed from $-k$ to k; the variable y is summed from $\max(-k, x - k)$ to $\min(k, x + k)$; and the variable z is summed from max $(-k, x - k, y - k)$ to min $(k, x + k, y + k)$. Then express the triple sum as the sum of six sums specified according to the relative sizes of $0, x$ and y.

3. First use the fact that $w^2 \equiv -1 \pmod{p}$ to prove that there are integers a and c such that $p = a^2 + c^2$. Then let s and t be integers such that $at - cs = 1$. Prove that $as + ct \equiv fw \pmod{p}$, where $f = \pm 1$, and deduce that an integer g can be found so that $b \ (= s - ag)$ and $d \ (= t - cg)$ satisfy $ab + cd = fw$, $ad - bc = 1$ and $b^2 + d^2 = (w^2 + 1)/p$.

4. Prove that

$$\sum_{n=1}^{m} (d_1(n) - d_3(n)) = \sum_{n=1}^{m} \sum_{\substack{d|n \\ d \text{ odd}}} (-1)^{(d-1)/2} \, ,$$

and then interchange the order of summation of the sums on the right side.

5. Rule out the possibilities $x \equiv 0 \pmod{2}$ and $x \equiv 3 \pmod{4}$ by congruence considerations. If $x \equiv 1 \pmod{4}$, prove that there is at least one prime $p \equiv 3 \pmod{4}$ dividing $x^2 - 3x + 9$. Deduce that p divides $x^3 + 27$, and then obtain a contradiction.

6. Use the identity

$$f(x_1, y_1)f(x_2, y_2) = (ax_1x_2 + bx_1y_2 + bx_2y_1 + cy_1y_2)^2 + (ac - b)^2(x_1y_2 - x_2y_1)^2$$

together with simple inequalities.

7. Prove that exactly one of the triples

$$(a, b, c) = (R, S, T), \ (T, -S + 2T, R - S + T), \ (R - S + T, 2R - S, R) \, ,$$

satisfies

$$a \le b < c, \quad \text{or} \quad a \ge b > c \, ,$$

by considering cases depending upon the relative sizes of R, S and T.

8. Consider the sign of the discriminant of
$$(aB - bA)x^2 + 2(aC - cA)xy + (bC - cB)y^2 .$$

9. Prove that the quantity
$$\left| \frac{n}{2^n} \sum_{k=1}^{n} \frac{2^k}{k} - \sum_{k=0}^{n-1} \frac{1}{2^k} \right|$$
tends to zero as $n \to \infty$.

10. Consider the case when $n = p + 1$ and $m = p$, where p is a prime suitably large compared with c.

11. Assume that $2D(D + A_1 A_2 + \epsilon B_1 B_2)$ is a square, where $\epsilon = \pm 1$. If D is odd, show that
$$\begin{cases} D + A_1 A_2 + \epsilon B_1 B_2 = 2DU^2 , \\ D - A_1 A_2 - \epsilon B_1 B_2 = 2DV^2 , \\ A_1 B_2 - \epsilon A_2 B_1 = 2DUV . \end{cases}$$
Deduce that $U^2 + V^2 = 1$. Then consider the four possibilities $(U, V) = (\pm 1, 0)$, $(0, \pm 1)$. The case D even can be treated similarly.

12. Set
$$\alpha_\pm = \sqrt{1985 \pm 31\sqrt{1985}}, \quad \beta_\pm = \sqrt{3970 \pm 64\sqrt{1985}} ,$$
and prove that
$$\alpha_+ + \alpha_- = \beta_+, \quad \alpha_+ - \alpha_- = \beta_- .$$

13.   If (x, y) is a solution of (13.0), prove that

$$x^2 + xy - y^2 = \pm F \ ,$$

and then solve the system of equations

$$\begin{cases} x^2 & + y^2 = k \ , \\ x^2 +2xy +2y^2 = l \ , \\ x^2 + xy - y^2 = \pm F \ , \end{cases}$$

for x^2, xy and y^2.

14.   Factor the left side of (14.0).

15.   Make the following argument mathematically rigorous:

$$\int_0^1 \ln x \ \ln(1 - x) \ dx = -\int_0^1 \ln x \ \sum_{k=1}^{\infty} \frac{x^k}{k} \ dx$$

$$= -\sum_{k=1}^{\infty} \frac{1}{k} \int_0^1 x^k \ln x \ dx$$

$$= \sum_{k=1}^{\infty} \frac{1}{k(k + 1)^2}$$

$$= \sum_{k=1}^{\infty} \frac{1}{k(k + 1)} - \sum_{k=1}^{\infty} \frac{1}{(k + 1)^2}$$

$$= 1 - \left(\frac{\pi^2}{6} - 1 \right)$$

$$= 2 - \frac{\pi^2}{6} \ .$$

16. Taking $n = 1, 2, \ldots, 6$ in (16.0), we obtain

$$a(1) = 1/2, \quad a(2) = -1/3, \quad a(3) = 1/4,$$
$$a(4) = -1/5, \quad a(5) = 1/6, \quad a(6) = -1/7 \, .$$

This suggests that $a(n) = (-1)^{n+1}/(n+1)$, which can be proved by induction on n.

17. Consider three cases according to the following values of the Legendre symbol:

$$\left(\frac{n^2 + k}{p}\right) = 1 \quad \text{or} \quad \left(\frac{(n+1)^2 + k}{p}\right) = 1$$
$$\text{or} \quad \left(\frac{n^2 + k}{p}\right) = \left(\frac{(n+1)^2 + k}{p}\right) = -1 \, .$$

In the third case, the identity

$$(n^2 + n + k)^2 + k = (n^2 + k)\left((n+1)^2 + k\right)$$

is useful.

18. Rearrange the terms of the partial sum

$$\sum_{n=0}^{N} \left(\frac{1}{4n+1} + \frac{1}{4n+3} - \frac{1}{2n+2}\right) \, ,$$

and then let $N \to \infty$.

19. Use

$$\left(\frac{A_n}{n}\right)^2 = \left(a_n + \frac{A_n}{n} - a_n\right)^2 \leq 2a_n^2 + 2\left(\frac{A_n}{n} - a_n\right)^2$$

to prove that

$$\sum_{n=1}^{m} \left(\frac{A_n}{n}\right)^2 \leq 4 \sum_{n=1}^{m} a_n^2 + 2 \sum_{n=1}^{m} \left(\frac{A_n}{n}\right)^2 - 4 \sum_{n=1}^{m} \frac{a_n A_n}{n} .$$

Then use

$$-2a_n A_n = -(A_n^2 - A_{n-1}^2) - a_n^2 \leq -(A_n^2 - A_{n-1}^2)$$

to prove that

$$-2 \sum_{n=1}^{m} \frac{a_n A_n}{n} \leq - \sum_{n=1}^{m} \frac{A_n^2}{n(n+1)} .$$

Putting these two inequalities together, deduce that

$$\sum_{n=1}^{m} \left(1 - \frac{2}{n+1}\right) \left(\frac{A_n}{n}\right)^2 \leq 4 \sum_{n=1}^{m} a_n^2 .$$

20.   Use the identity

$$\frac{\binom{n}{k}}{\binom{2n-1}{k}} = 2 \left(\frac{\binom{n}{k}}{\binom{2n}{k}} - \frac{\binom{n}{k+1}}{\binom{2n}{k+1}} \right) .$$

21.   All integral solutions of $ax + by = k$ are given by

$$x = g + bt, \quad y = h - at, \quad t = 0, \pm 1, \pm 2, \dots ,$$

where (g, h) is a particular solution of $ax + by = k$.

22.   Prove that

$$u_k = v_{k-1} - v_k, \qquad k = 0, 1, \dots ,$$

where

$$v_k = \frac{1}{(a + (k+1)d)\cdots(a + (k+r)d)rd}, \qquad k = -1, 0, 1, \ldots .$$

23.   Prove that the stronger inequality

$$F(x_{12}, x_{13}, \ldots, x_{n-1\,n}) \leq \sum_{k=1}^{n} x_k^2 - \frac{1}{n}\left(\sum_{k=1}^{n} x_k\right)^2$$

holds by replacing each x_i by $x_i - M$ for suitable $M = M(x_1, \ldots, x_n)$ in (23.0).

24.   Apply the Cauchy-Schwarz inequality to

$$\sum_{n=1}^{m} a_n \sqrt{n}\, \frac{1}{\sqrt{n}} .$$

25.   Consider the integers $(3m + 2)^2$, $m = 1, 2, \ldots .$

26.   Use the identity

$$\arctan\left(\frac{2}{n^2}\right) = \arctan\left(\frac{1}{n-1}\right) - \arctan\left(\frac{1}{n+1}\right), \qquad n = 2, 3, \ldots .$$

27.   Suppose that $g_n(x) = h(x)k(x)$, where $h(x)$ and $k(x)$ are non-constant polynomials with integral coefficients. Show that $h(x)$ and $k(x)$ can be taken to be positive for all real x, and that $h(p_i) = k(p_i) = 1$, $i = 1, 2, \ldots, n$. Deduce that $h(x)$ and $k(x)$ are both of degree n, and determine the form of both $h(x)$ and $k(x)$. Obtain a contradiction by equating appropriate coefficients in $g_n(x)$ and $h(x)k(x)$.

28.   Let $p(k)$, $k = 0, 1, \dots$, denote the probability that A wins when A has k dollars. Prove the recurrence relation

$$ap(k+2) - (a+b)p(k+1) + bp(k) = 0.$$

29.   If $x_1, \dots, x_n$ are the n roots of $f(x)$, the $2n$ roots of $f(x^2)$ are ·

$$\pm\sqrt{x_1}, \pm\sqrt{x_2}, \dots, \pm\sqrt{x_n}.$$

30.   Find a point P such that any two different lattice point must be at different distances from P. Then consider the lattice points sequentially according to their increasing distances from P.

31.   Determine the generating function

$$\sum_{n=0}^{\infty} S(n) t^n.$$

32.   As $f(x)$ is bounded on $(0, a)$ there exists a positive constant K such that

$$|f(x)| < K, \quad 0 < x < a.$$

Use (32.0) to deduce successively that

$$\begin{cases} |f(x)| < K/n, & 0 < x < a, \\ |f(x)| < K/n^2, & 0 < x < a, \\ \dots \\ \text{etc}. \end{cases}$$

33. Consider the derivative of the function

$$f(x) = \frac{(y_1 - y_2)(y_3 - z)}{(y_1 - y_3)(y_2 - z)} .$$

34. Set $a_1 = 1$. Prove the recurrence relation

$$a_{n+1} = a_1 a_n + a_2 a_{n-1} + \cdots + a_{n-1} a_2 + a_n a_1 ,$$

and use it to show that the generating function $A(x) = \sum_{n=1}^{\infty} a_n x^n$ satisfies $A(x)^2 = A(x) - x$. Then solve for $A(x)$.

35. Use L'Hôpital's rule, or use the inequality

$$t \le \tan t \le t + t^3, \qquad 0 \le t \le 1 ,$$

to estimate the integral $\int_0^\pi \tan(y \sin x) \, dx$.

36. Use a result due to Hurwitz, namely, if θ is an irrational number, there are infinitely many rational numbers a/b with $b > 0$ and $GCD(a, b) = 1$ such that

$$|\theta - a/b| < 1/(\sqrt{5}\, b^2) .$$

37. Differentiate (37.0) with respect to x and y to obtain

$$(1 + x^2)f'(x) = (1 + y^2)f'(y) .$$

38.   Introduce a coordinate system and use simple inequalities to show that $\max S = 4R$ and $\min S = 2R$, where R is the radius of the circle.

39.   Prove that

$$\bigcup_{n=1}^{\infty} \left(\bigcap_{k=1}^{\infty} A_{n+k} \right) = X \cap Y$$

and

$$\bigcap_{n=1}^{\infty} \left(\bigcup_{k=1}^{\infty} A_{n+k} \right) = X \cup Y \,,$$

where

$$X = \{\, 0, 2, 4, \dots \,\}, \quad Y = \{\, 0, 3, 6, \dots \,\} \,.$$

40.   Prove that

$$p_k = \begin{cases} 0 & ,\ 0 \le k \le n-1 \,, \\ p^n & ,\ k = n \,, \\ qp^n & ,\ n+1 \le k \le 2n \,, \end{cases}$$

and

$$p_k = \left(1 - \sum_{i=0}^{k-n-1} p_i \right) qp^n, \quad k > 2n \,.$$

Use these to find a linear equation satisfied by $P(x)$.

41.   First prove that the circumradius of a triangle with sides of length l, m and n is given by

$$\frac{lmn}{\sqrt{(l+m+n)(l+m-n)(l-m+n)(-l+m+n)}} \,.$$

Next show that

$$|AC| = \sqrt{\frac{(ac+bd)(ad+bc)}{(ab+cd)}} \,.$$

Finally, apply the above two results to $\triangle ABC$.

42. Consider the quadrilateral $ABCD$ as lying in the complex plane. Represent the vertices A, B, C, D by the complex numbers a, b, c, d respectively. Prove that P, Q, R, S are represented by the numbers

$$\left(\tfrac{1-i}{2}\right)(a+ib), \qquad \left(\tfrac{1-i}{2}\right)(b+ic),$$
$$\left(\tfrac{1-i}{2}\right)(c+id), \qquad \left(\tfrac{1-i}{2}\right)(d+ia),$$

respectively. Then relate $p - r$ and $q - s$.

43. Try a solution of the form

$$p = xy + X, \qquad q = y + Y,$$

where X and Y are polynomials in x, w and z. Substitute in (43.0) and solve the resulting equations for X and Y.

44. Set $\alpha = f(1)$ and $\beta = f(i)$. Prove that $|\alpha| = |\beta| = 1, |\alpha - \beta| = \sqrt{2}$. Deduce that $\alpha^2 + \beta^2 = 0$ so that $\beta = \epsilon\alpha, \ \epsilon = \pm i$. Next from (44.0) deduce that

$$\begin{cases} \overline{\alpha}f(z) + \alpha\overline{f(z)} &= z + \overline{z}, \\ \overline{\alpha}f(z) - \alpha\overline{f(z)} &= -\epsilon iz + \epsilon i\overline{z}. \end{cases}$$

Now solve for $f(z)$.

45. Let x be a rational number such that $y = \tan \pi x$ is rational. Prove that $z = 2\cos 2\pi x$ is a rational root of a monic polynomial with integral coefficients. Deduce that $z = 0, \pm 1, \pm 2$.

46. Let S_1, S_2, S_3 denote the areas of $\triangle PBC, \triangle PCA, \triangle PAB$ respectively. Prove that

$$\frac{|PA|}{|PD|} = \frac{S_2 + S_3}{S_1}$$

with similar expressions for $\frac{|PB|}{|PE|}$ and $\frac{|PC|}{|PF|}$.

47.   Prove that the (i,j)-th entry of N is l times the (i,j)-th entry of M modulo n.

48.   There are $(N+1)^n$ vectors $(y_1, y_2, \ldots, y_n)$ of integers satisfying $0 \leq y_j < N$, $1 \leq j \leq n$. For each of these vectors the corresponding value of

$$L_i = L_i(y_1, y_2, \ldots, y_n) = a_{i1}y_1 + \cdots + a_{in}y_n, \ 1 \leq i \leq m ,$$

satisfies $-naN \leq L_i \leq naN$, so the vector $(L_1, L_2, \ldots, L_m)$ of integers can take on at most $(2naN + 1)^m$ different values. Choose N appropriately and apply Dirichlet's box principle.

49.   Suppose that $\int e^{-x^2} dx$ is an elementary function, so that by Liouville's result, there is a rational function $p(x)/q(x)$, where $p(x)$ and $q(x)$ are polynomials with no common factor, such that

$$\int e^{-x^2} \, dx = \frac{p(x)}{q(x)} e^{-x^2} .$$

Differentiate both sides to obtain

$$p'(x)q(x) - p(x)q'(x) - 2xp(x)q(x) = q(x)^2 ,$$

and deduce that $q(x)$ is a nonconstant polynomial. Let c denote one of the complex roots of $q(x)$ and obtain a contradiction by expressing $q(x)$ in the form $q(x) = (x-c)^m r(x)$, with $r(x)$ not divisible by $(x-c)$.

50.   Prove that

$$x_n = \sum_{i=0}^{n-1} \frac{(-1)^i}{i+1}, \quad n = 1, 2, \ldots .$$

51. Let p_k denote the k-th prime. Suppose that $N > 121$ is an integer with the property (51.0). Let p_n be the largest prime less than or equal to $\sqrt{N}$, so that $n \geq 5$, and $N < p_{n+1}^2$. Use property (51.0) to obtain the inequality $N \geq p_1 p_2 \cdots p_n$. Then use Bertrand's postulate

$$p_{k+1} \leq 2p_k, \ k = 1, 2, \ldots ,$$

to obtain

$$p_1 p_2 \cdots p_{n-2} < 8$$

from the inequality $p_1 p_2 \cdots p_n < p_{n+1}^2$. Deduce the contradiction $n \leq 4$. Check property (51.0) for the integers $N = 3, 4, \ldots, 121$ directly.

52. Prove that

$$S = \int_0^1 \frac{x^2 + x + 1}{x^4 + x^3 + x^2 + x + 1} \, dx$$

and then use partial fractions to evaluate the integral.

53. Let $|AB| = 2c$, $|BC| = 2a$, $|CA| = 2b$. Show that

$$l = \sqrt{(a - b + c)(a + b - c)}$$

with similar expressions for m and n.

54. Evaluate

$$H(1,1,0,0), \quad H(0,0,1,1),$$
$$H(0,1,1,0), \quad H(1,0,0,1),$$

using (i) and (iii). Then express $H(a,b,c,d)$ in terms of these quantities by means of $(i), (ii), (iii)$ and (iv).

55. Set $f(z) = z^n + a_1 z^{n-1} + \cdots + a_n$ and note that for $z \neq 0$ we have

$$|f(z)| = \left| z^n \left(1 + \frac{a_1}{z} + \cdots + \frac{a_n}{z^n} \right) \right|$$

$$= \quad |z^n| \left| 1 + \frac{a_1}{z} + \cdots + \frac{a_n}{z^n} \right|$$

$$\geq \quad |z^n| \left(1 - \frac{|a_1|}{|z|} - \cdots - \frac{|a_n|}{|z|^n} \right)$$

$$\geq \quad |z^n| \left(1 - \frac{A}{|z|} - \cdots - \frac{A}{|z|^n} \right) \ .$$

56.   Define integers k and l by

$$2^m - 1 = kd, \quad 2^n + 1 = ld \ ,$$

and then consider

$$2^{mn} = (kd + 1)^n = (ld - 1)^m \ .$$

57.   Suppose that $f(x) = g(x)h(x)$, where $g(x)$ and $h(x)$ are nonconstant polynomials with integral coefficients chosen so that

$$\deg(g(x)) \leq \deg(h(x)) \ .$$

Deduce that $\deg(g(x)) \leq m$ and that $g(k_i) = \pm 1$, $i = 1, 2, \ldots, 2m + 1$. Let $\epsilon = +1$ (resp. -1) if $+1$ (resp. -1) occurs at least $m + 1$ times among the values $g(k_i) = \pm 1$, $i = 1, 2, \ldots, 2m + 1$. Then consider the polynomial $g(x) - \epsilon$.

58.   Suppose a, b, c, d are integers, not all zero, satisfying (58.0). Show that without loss of generality a, b, c, d may be taken to satisfy

$$GCD(a, b, c, d) = 1 \ .$$

By considering (58.0) modulo 5 prove that

$$a \equiv b \equiv c \equiv d \equiv 0 \ (\text{mod } 5) \ .$$

59.   Consider the integers $72k + 42$, $k = 0, 1, \dots$.

60.   Use the identity

$$2^k \sin kx \, \cos^k x = \sum_{r=1}^{k} \binom{k}{r} \sin 2rx \ .$$

61.   Express

$$\frac{1}{n+1} \binom{2n}{n}$$

as the difference of two binomial coefficients.

62.   Use the identity

$$\frac{2^n}{a^{2^n} + 1} = \frac{2^n}{a^{2^n} - 1} - \frac{2^{n+1}}{a^{2^{n+1}} - 1} \ , \quad a > 1 \ .$$

63.   Prove that

$$a_n = 2(-1)^{n-1} \frac{1}{n} \binom{2n-2}{n-1} \left(\frac{k}{4}\right)^n$$

and appeal to Problem 61.

64.   Let $C_1, C_2, \dots, C_m$ denote the columns of M_m. Determine a linear combination of $C_1, C_2, \dots, C_{m-1}$ which when added to C_m gives the column $(0, 0, \dots, \phi(m))^t$. Deduce that $\det M_m = \phi(m) \det M_{m-1}$.

65.   Assume (65.0) holds and use congruences modulo 8 to obtain a contradiction.

66.   Prove that

$$a_n = \int_0^1 \frac{(-1)^{n-1} x^n}{1+x} \, dx \ ,$$

and use this representation of a_n to deduce that

$$\left| \sum_{n=1}^{N} a_n - \int_0^1 \frac{x}{(1+x)^2} \, dx \right| \le \frac{1}{N+2} \ .$$

67.   Let A be a sequence of the required type, and let k denote the number of zeros in A. First prove that $k = 3$. Deduce that $A = \{0, 0, 0, a_3, a_4, a_5, 3\}$, where $1 \le a_3 \le a_4 \le a_5 \le 3$. Then prove that $N_1 = 2$.

68.   Prove that

$$g^n h g^{-n} = h^{2^n}, \quad n = 1, 2, \ldots, 5 \ .$$

69.   Prove that every integer $N > 2ab$ is of the form

$$N = xa + yb, \quad x \ge 0, \quad y \ge 0, \quad x + y \equiv 1 \, (\mathrm{mod}\ 2) \ ,$$

and that all integers of this form belong to S.

70.   If m is even, say $m = 2n$, show that

$$m = (an + b)^2 + (cn + d)^2 - 5(en + f)^2 \ ,$$

for suitable constants $a, b, \ldots, f$. The case m odd is treated similarly.

71. Note that

$$\frac{\ln 2}{2} - \frac{\ln 3}{3} + \frac{\ln 4}{4} - \cdots + \frac{\ln 2n}{2n}$$
$$= \ln 2 \left(1 + \frac{1}{2} + \cdots + \frac{1}{n}\right) + \sum_{k=1}^{n} \frac{\ln k}{k} - \sum_{k=1}^{2n} \frac{\ln k}{k} ,$$

and estimate $\sum_{k=1}^{n} (\ln k)/k$ for large n using the Euler-MacLaurin summation formula.

72. Express $(\sqrt{k+1} - \sqrt{k})^3$ in the form $\sqrt{p(k)+1} - \sqrt{p(k)}$, where $p(k)$ is a cubic polynomial in k.

73. Choose a_1 to be any nonzero integer such that $a_1 \equiv a \pmod{n}$. Then set $b_1 = b + rn$, where r is the product of those primes which divide a_1 but do not divide either b or n. Prove that $GCD(a_1, b_1) = 1$.

74. Prove that $s(n+1) \equiv 2s(n) \pmod{3}$, and use this conguence to show that there are no positive integers n satisfying $s(n) = s(n+1)$.

75. Show that

$$S = \sum_{m,n=1}^{\infty} \frac{1}{mn(m+n)} \bigg/ \sum_{d=1}^{\infty} \frac{1}{d^3}$$

by collecting together those m, n in the sum $A = \sum_{m,n=1}^{\infty} 1/(mn(m+n))$ having the same value for $GCD(m, n)$. Then evaluate the sum A by proving that it is equal to the integral

$$\int_0^1 \frac{\ln^2(1-x)}{x} \, dx ,$$

which can be evaluated by means of the transformation $x = 1 - e^{-u}$.

76.   Apply the intermediate value theorem to the function $T(x)$ defined to be the time taken in minutes by the racer to run from the point x miles along the course to the point $x + 2$ miles along the course.

77.   Choose a coordinate system so that

$$A = (-1,0), \quad O = (0,0), \quad B = (1,0) .$$

Then $R = (-a,0)$ with $0 < a < 1$. Deduce that

$$
\begin{aligned}
C &= (1 - 2a, 0) , \\
S &= (1 - a, 0) , \\
D &= \left(1 - 2a, 2\sqrt{a(1 - a)}\right) , \\
P &= \left(2a^2 - 4a + 1, 2(1 - a)\sqrt{a(1 - a)}\right) , \\
Q &= \left(1 - 2a^2, 2a\sqrt{a(1 - a)}\right) ,
\end{aligned}
$$

and calculate the slopes of PC, PD, QC and QD.

78.   Let I denote the $n \times n$ identity matrix. Set $U = S + I$. Prove that $U^2 = nU$. Seek an inverse of S of the form $cU - I$.

79.   Replace $\cos(k\pi/n)$ by $(\omega^k + \omega^{-k})/2$, where $\omega = \exp(\pi i/n)$, and use the binomial theorem.

80.   Let $A = \begin{bmatrix} -1 & 1 \\ 0 & -2 \end{bmatrix}$ and show that $A^3 + 3A^2 + 2A = 0$. Then consider $(A + I)^3$.

81.   Let the sides of the triangles be a, b, c and b, c, d. The two triangles are similar if $a/b = b/c = c/d$. Choose positive integers to satisfy this relation remembering that the triangle inequalities $c < a + b$, etc must be satisfied.

82. Apply the mean value theorem to $f(x)$ on the intervals (a, b) and (b, c).

83. First show that $\lim_{n \to \infty} na_n = 0$. Then let $n \to \infty$ in

$$\sum_{k=1}^{n} k(a_k - a_{k+1}) = \sum_{k=1}^{n} a_k - na_{n+1} .$$

84. Use the fact that the length of the period of the continued fraction of $\sqrt{D}$ is one, and that D is an odd nonsquare integer > 5, to show that $D = 4c^2 + 1$, $c \geq 2$. Then determine the continued fraction of $\frac{1}{2}(1 + \sqrt{D})$.

85. For $x, y \in G$ prove that $(xyx^{-1}y^{-1})^2 = 1$.

86. Let λ denote one of the eigenvalues of A and let $\underline{x}$ be a nonzero eigenvector of A corresponding to λ. By applying simple inequalities to an appropriate row of $A\underline{x} = \lambda\underline{x}$, deduce that $|\lambda - 1| \leq 1$. Then use the fact that A is real symmetric and the relationship between $\det A$ and the eigenvalues of A.

87. Show that there exists an integer $k \geq 2$ such that $r = r^k$. Then prove that r^{k-1} is a multiplicative identity for R.

88. For $1 \leq k \leq l \leq n$ let $S(k, l)$ denote the set of subsets of J_n with $\min_{s \in S} S = k$ and $\max_{s \in S} S = l$. Evaluate $|S(k, l)|$ and then compute

$\sum_{\phi \neq S \subseteq J_n} w(S)$ using

$$\sum_{\phi \neq S \subseteq J_n} w(S) = \sum_{1 \leq k < l \leq n} (l - k)|S(k,l)| .$$

89.   Write n in binary notation, say,

$$n = 2^{a_1} + 2^{a_2} + \cdots + 2^{a_k} ,$$

where $a_1, \ldots, a_k$ are integers such that $a_1 > a_2 > \cdots > a_k \geq 0$, and then use

$$(1 + x)^{2^a} \equiv 1 + x^{2^a} \pmod 2 ,$$
$$(1 + x)^n = (1 + x)^{2^{a_1}}(1 + x)^{2^{a_2}} \cdots (1 + x)^{2^{a_k}} .$$

90.   Suppose that x_i, $1 \leq i \leq n$ belongs to the r_i-th row and the s_i-th column. Show that

$$\sum_{i=1}^{n} x_i = n \sum_{i=1}^{n} r_i - n^2 + \sum_{i=1}^{n} s_i ,$$

and then use the fact that both $\{ r_1, \ldots, r_n \}$ and $\{ s_1, \ldots, s_n \}$ are permutations of $\{ 1, 2, \ldots, n \}$.

91.   Let $N_{xx}, N_{xo}, N_{ox}, N_{oo}$ denote the number of occurrences of XX, XO, OX, OO respectively. Relate $N_{xx}, N_{xo}, N_{ox}, N_{oo}$ to a, b, p, q. Prove that $N_{ox} = N_{xo}$, and deduce the value of $a - b$ in terms of p and q.

92.   Consider the entries of the triangular array modulo 2. Show that the pattern

$$
\begin{array}{cccc}
 & 1 & 1 & 0 & 1 \\
1 & 0 & 0 & 0 \\
1 & 1 & 1 & 0 \\
1 & 0 & 1 & 0
\end{array}
$$

is repeated down the left edge of the array from the fourth row down.

93. Let x_{n+1} be any real number such that $|x_{n+1}| = |x_n + c|$, and consider $\sum_{i=1}^{n+1} x_i^2$.

94. If we have $f(x) = g(x)h(x)$ then without loss of generality $g(0) = \pm 1$, $h(0) = \pm 5$. Prove that one of the complex roots β of $g(x)$ satisfies $|\beta| \le 1$, and then deduce that $|f(\beta)| \ge 1$.

95. Set
$$f(x) = (x - a_1)(x - a_2) \cdots (x - a_n) .$$
Prove that
$$\left(\frac{1}{f'(a_1)}, \ldots, \frac{1}{f'(a_n)} \right) \quad \text{and} \quad \left(\frac{a_1}{f'(a_1)}, \ldots, \frac{a_n}{f'(a_n)} \right)$$
are two solutions of (95.0). Deduce the general solution of (95.0) from these two solutions.

96. By picking out the terms with $n = N$ in the sum $s(N)$, show that $s(N) = s(N - 1)$ for $N \ge 3$.

97. Prove that
$$L = \int_0^1 \int_0^1 \frac{x}{x^2 + y^2} \, dx \, dy ,$$
and evaluate the double integral using polar coordinates.

98. For convenience set $p = \pi/11$, and let $c = \cos p$, $s = \sin p$. Use the imaginary part of
$$(c + is)^{11} = -1 ,$$

to prove that

$$(11s - 44s^3 + 32s^5)^2 = 11c^2(1 - 4s^2)^2 \ .$$

Then show that

$$\tan 3p + 4\sin 2p = \frac{11s - 44s^3 + 32s^5}{c(1 - 4s^2)} = \pm\sqrt{11} \ .$$

Deduce that the + sign holds by considering the sign of the left side.

99.   Use partial summation and the fact that $\lim_{k\to\infty}(c_k - \ln k)$ exists.

100.   Use the identity

$$\frac{1}{(x-1)} \frac{x^{2^n}}{(x+1)(x^2+1)(x^4+1)\cdots(x^{2^n}+1)}$$
$$= \frac{x^{2^n}}{(x^{2^{n+1}}-1)} = \frac{1}{x^{2^n}-1} - \frac{1}{x^{2^{n+1}}-1} \ .$$

THE SOLUTIONS

Some people think we are wrong, but only time will tell: given all the alternatives, we have the solution.

Lev Davydovich Bronstein Trotsky (1879-1940)

1. Let p denote an odd prime and set $\omega = \exp(2\pi i/p)$. Evaluate the product

$$(1.0) \quad E(p) = (\omega^{r_1} + \omega^{r_2} + \ldots + \omega^{r_{(p-1)/2}})(\omega^{n_1} + \omega^{n_2} + \ldots + \omega^{n_{(p-1)/2}}),$$

where $r_1, \ldots, r_{(p-1)/2}$ denote the $(p-1)/2$ quadratic residues modulo p and $n_1, \ldots, n_{(p-1)/2}$ denote the $(p-1)/2$ quadratic nonresidues modulo p.

Solution: We set $q = (p-1)/2$ and

$$(1.1) \quad \epsilon = \begin{cases} 0, & \text{if } p \equiv 1 \pmod 4, \\ 1, & \text{if } p \equiv 3 \pmod 4, \end{cases}$$

and for $k = 0, 1, \ldots, p-1$ let

$$(1.2) \quad N(k) = \sum_{\substack{i,j=1 \\ r_i + n_j \equiv k \pmod p}}^{q} 1.$$

If k is a quadratic residue (resp. nonresidue) (mod p) $\{kr_i : i = 1, 2, \ldots, q\}$ is a complete system of quadratic residues (resp. nonresidues) (mod p) and $\{kn_j : j = 1, 2, \ldots, q\}$ is a complete system of quadratic nonresidues (resp. residues) (mod p). Replacing r_i by kr_i and n_j by kn_j in (1.2), where $1 \leq k \leq p - 1$, we obtain

$$(1.3) \qquad N(k) = N(1), \quad k = 1, 2, \ldots, p - 1 \, .$$

Next, we note that

$$(1.4) \qquad N(0) = \sum_{\substack{i,j=1 \\ r_i \equiv -n_j \,(\mathrm{mod}\, p)}}^{q} 1 \; = \epsilon q \, ,$$

as -1 is a quadratic residue (mod p) for $p \equiv 1 \pmod 4$ and -1 is a quadratic nonresidue (mod p) for $p \equiv 3 \pmod 4$. Now as

$$(1.5) \qquad \sum_{k=0}^{p-1} N(k) = \sum_{i,j=1}^{q} 1 = q^2 \, ,$$

we obtain, from (1.3), (1.4), and (1.5),

$$\epsilon q + 2q N(1) = q^2 \, ,$$

that is

$$(1.6) \qquad N(1) = (q - \epsilon)/2 \, .$$

Finally, we have

$$
\begin{aligned}
E(p) &= \left(\sum_{i=1}^{q} \omega^{r_i} \right) \left(\sum_{j=1}^{q} \omega^{n_j} \right) \\
&= \sum_{i,j=1}^{q} \omega^{r_i + n_j}
\end{aligned}
$$

$$= \sum_{k=0}^{p-1} \sum_{\substack{i,j=1 \\ r_i+n_j \equiv k \pmod{p}}}^{q} \omega^{r_i+n_j}$$

$$= \sum_{k=0}^{p-1} \omega^k N(k)$$

$$= N(0) + N(1)(\omega + \omega^2 + \cdots + \omega^{p-1})$$

$$= N(0) - N(1)$$

$$= \epsilon q - (q - \epsilon)/2, \quad \text{by (1.4) and (1.6)},$$

that is

$$E(p) = \begin{cases} (1-p)/4, & \text{if } p \equiv 1 \pmod{4}, \\ (1+p)/4, & \text{if } p \equiv 3 \pmod{4}, \end{cases}$$

as required.

2. Let k denote a positive integer. Determine the number $N(k)$ of triples (x, y, z) of integers satisfying

$$(2.0) \qquad \begin{cases} |x| \le k, & |y| \le k, & |z| \le k, \\ |x - y| \le k, & |y - z| \le k, & |z - x| \le k. \end{cases}$$

Solution: The required number $N(k)$ of triples is given by

$$N(k) = \sum_{\substack{|x| \le k \\ |x-y| \le k}} \sum_{\substack{|y| \le k \\ |y-z| \le k \\ |z-x| \le k}} \sum_{|z| \le k} 1$$

$$= \sum_{x=-k}^{k} \sum_{\substack{y=-k \\ x-k \le y \le x+k}}^{k} \sum_{\substack{z=-k \\ x-k \le z \le x+k \\ y-k \le z \le y+k}}^{k} 1,$$

that is

$$(2.1) \qquad N(k) = \sum_{x=-k}^{k} \sum_{y} \sum_{z} 1 \, ,$$

where the second sum is taken over $y = \max\left(-k, x - k\right)$ to $y = \min\left(k, x + k\right)$, and the third sum is taken over $z = \max\left(-k, x - k, y - k\right)$ to $z = \min\left(k, x + k, y + k\right)$. We now split the sum on the right of (2.1) into six sums $S_1, \ldots, S_6$, where x and y are restricted as follows:

$$
\begin{aligned}
0 \le x \le y \, , && \text{in } S_1 \, ; \\
x < 0 \le y \, , && \text{in } S_2 \, ; \\
x \le y < 0 \, , && \text{in } S_3 \, ; \\
0 \le y < x \, , && \text{in } S_4 \, ; \\
y < 0 \le x \, , && \text{in } S_5 \, ; \\
y < x < 0 \, , && \text{in } S_6 \, .
\end{aligned}
$$

Clearly, we have

$$
\begin{aligned}
S_1 &= \sum_{x=0}^{k} \sum_{y=x}^{k} \sum_{z=y-k}^{k} 1 \\
&= \sum_{x=0}^{k} \sum_{y=x}^{k} (2k + 1 - y) \\
&= \frac{1}{2} \sum_{x=0}^{k} (k + 1 - x)(3k + 2 - x) \\
&= \frac{1}{2} \sum_{x=0}^{k} \left((k + 1)(3k + 2) - (4k + 3)x + x^2 \right) \\
&= \frac{1}{2} \left((k + 1)^2(3k + 2) - \frac{(4k + 3)k(k + 1)}{2} + \frac{k(k + 1)(2k + 1)}{6} \right) \\
&= \frac{1}{6}(k + 1)(k + 2)(4k + 3) \, .
\end{aligned}
$$

Similarly, with E denoting $k(k + 1)(2k + 1)/3$, we obtain

$$
S_2 = \sum_{x=-k}^{-1} \sum_{y=0}^{x+k} \sum_{z=y-k}^{x+k} 1 \quad = E \, ,
$$

$$S_3 = \sum_{x=-k}^{-1} \sum_{y=x}^{-1} \sum_{z=-k}^{x+k} 1 = E,$$

$$S_4 = \sum_{x=1}^{k} \sum_{y=0}^{x-1} \sum_{z=x-k}^{k} 1 = E,$$

$$S_5 = \sum_{x=0}^{k} \sum_{y=x-k}^{-1} \sum_{z=x-k}^{y+k} 1 = E,$$

$$S_6 = \sum_{x=-k+1}^{-1} \sum_{y=-k}^{x-1} \sum_{z=-k}^{y+k} 1 = \frac{1}{6}(k-1)k(4k+1).$$

Thus we have

$$
\begin{aligned}
N(k) &= S_1 + S_2 + \cdots + S_6 \\
&= \frac{1}{6}(k+1)(k+2)(4k+3) + \frac{4}{3}k(k+1)(2k+1) \\
&\quad + \frac{1}{6}(k-1)k(4k+1) \\
&= 4k^3 + 6k^2 + 4k + 1 \\
&= (k+1)^4 - k^4 .
\end{aligned}
$$

3. Let $p \equiv 1 \pmod 4$ be prime. It is known that there exists a unique integer $w \equiv w(p)$ such that

$$w^2 \equiv -1 \pmod p, \quad 0 < w < p/2.$$

(For example, $w(5) = 2, w(13) = 5$.) Prove that there exist integers a, b, c, d with $ad - bc = 1$ such that

$$pX^2 + 2wXY + \frac{(w^2+1)}{p}Y^2 \equiv (aX + bY)^2 + (cX + dY)^2.$$

(For example, when $p = 5$ we have

$$5X^2 + 4XY + Y^2 \equiv X^2 + (2X + Y)^2,$$

and when $p = 13$ we have

$$13X^2 + 10XY + 2Y^2 \equiv (3X + Y)^2 + (2X + Y)^2.)$$

Solution: We make use of the following property of the reals: if r is any real number, and n is a positive integer, then there exists a rational number h/k such that

$$(3.1) \qquad \left| r - \frac{h}{k} \right| \leq \frac{1}{k(n+1)}, \quad 1 \leq k \leq n, \quad GCD(h,k) = 1 .$$

Taking $r = -w(p)/p$ and $n = [\sqrt{p}]$, we see that there are integers a and e such that

$$(3.2) \qquad \left| \frac{-w(p)}{p} - \frac{e}{a} \right| < \frac{1}{a\sqrt{p}}, \quad 1 \leq a < \sqrt{p} .$$

Setting $c = w(p)a + pe$, we see from (3.2) that $|c| < \sqrt{p}$, and so $0 < a^2 + c^2 < 2p$. But $c \equiv wa \pmod{p}$, and so $a^2 + c^2 \equiv a^2(1 + w^2) \equiv 0 \pmod{p}$, showing that

$$(3.3) \qquad p = a^2 + c^2 .$$

As p is a prime, we see from (3.3) that $GCD(a,c) = 1$. Hence, we can choose integers s and t such that

$$(3.4) \qquad at - cs = 1 .$$

Hence

$$
\begin{aligned}
(as &+ ct - w)(as + ct + w) \\
&= (as + ct)^2 - w^2 \\
&= (a^2 + c^2)(s^2 + t^2) - (at - cs)^2 - w^2 \\
&= p(s^2 + t^2) - (1 + w^2) \\
&\equiv 0 \pmod{p} ,
\end{aligned}
$$

so that

$$(3.5) \qquad as + ct \equiv fw \pmod{p}, \quad f = \pm 1 .$$

Hence there is an integer g such that

(3.6) $$as + ct = fw + gp .$$

Set

(3.7) $$b = s - ag, \quad d = t - cg .$$

Then, by (3.3), (3.4), (3.6), and (3.7), we have

(3.8) $$ab + cd = fw, \quad ad - bc = 1 .$$

We now obtain

$$
\begin{aligned}
p(b^2 + d^2) &= (a^2 + c^2)(b^2 + d^2) \\
&= (ab + cd)^2 + (ad - bc)^2 \\
&= w^2 + 1 ,
\end{aligned}
$$

so that

(3.9) $$b^2 + d^2 = (w^2 + 1)/p .$$

Then, from (3.3), (3.8), and (3.9), we have

(3.10) $$(aX + bY)^2 + (cX + dY)^2 = pX^2 + 2fwXY + \frac{(w^2 + 1)}{p}Y^2 .$$

If $f = 1$ then (3.10) is the required identity. If $f = -1$, replace b, c, Y by $-b, -c, -Y$ respectively to obtain the desired result.

4. Let $d_r(n)$, $r = 0, 1, 2, 3$, denote the number of positive integral divisors of n which are of the form $4k + r$. Let m denote a positive integer. Prove that

(4.0) $$\sum_{n=1}^{m}(d_1(n) - d_3(n)) = \sum_{j=0}^{\infty}(-1)^j \left[\frac{m}{2j+1} \right] .$$

Solution: We have

$$
\sum_{n=1}^{m}(d_1(n) - d_3(n)) = \sum_{n=1}^{m} \sum_{\substack{d \mid n \\ d \text{ odd}}} (-1)^{(d-1)/2}
$$

$$
= \sum_{d \text{ odd}} \sum_{1 \le dk \le m} (-1)^{(d-1)/2}
$$

$$
= \sum_{d \text{ odd}} (-1)^{(d-1)/2} \sum_{1 \le k \le m/d} 1
$$

$$
= \sum_{d \text{ odd}} (-1)^{(d-1)/2} \left[\frac{m}{d} \right]
$$

$$
= \sum_{j=0}^{\infty} (-1)^j \left[\frac{m}{2j+1} \right].
$$

This completes the proof of (4.0).

     5.   Prove that the equation

(5.0) $$y^2 = x^3 + 23$$

has no solutions in integers x and y.

Solution: Suppose that (x, y) is a solution of (5.0) in integers. If $x \equiv 0 \pmod 2$ then (5.0) gives $y^2 \equiv 3 \pmod 4$, which is impossible. Hence, we must have $x \equiv 1 \pmod 2$. If $x \equiv 3 \pmod 4$ then (5.0) gives $y^2 \equiv 2 \pmod 4$, which is impossible. Hence, we see that $x \equiv 1 \pmod 4$. In this case we have $x^2 - 3x + 9 \equiv 3 \pmod 4$, and so there is at least one prime $p \equiv 3 \pmod 4$ dividing $x^2 - 3x + 9$. Since $x^2 - 3x + 9$ is a factor of $x^3 + 27$, we have $x^3 + 27 \equiv 0 \pmod p$. Thus by (5.0) we have $y^2 \equiv -4 \pmod p$. This congruence is insolvable as -4 is not a quadratic residue for any prime $p \equiv 3 \pmod 4$, showing that (5.0) has no solutions in integers x and y.

     6.   Let $f(x, y) = ax^2 + 2bxy + cy^2$ be a positive-definite quadratic form.

Prove that

(6.0)
$$(f(x_1, y_1)f(x_2, y_2))^{1/2}f(x_1 - x_2, y_1 - y_2)$$
$$\geq (ac - b^2)(x_1 y_2 - x_2 y_1)^2,$$

for all real numbers x_1, x_2, y_1, y_2.

Solution: First we note that $ac - b^2 > 0$ as f is positive-definite. We use the identity

(6.1)
$$(ax_1^2 + 2bx_1y_1 + cy_1^2)(ax_2^2 + 2bx_2y_2 + cy_2^2) =$$
$$(ax_1x_2 + bx_1y_2 + bx_2y_1 + cy_1y_2)^2 + (ac - b^2)(x_1y_2 - x_2y_1)^2.$$

Set
$$E_1 = f(x_1, y_1) \geq 0, \quad E_2 = f(x_2, y_2) \geq 0,$$
$$F = |ax_1x_2 + bx_1y_2 + bx_2y_1 + cy_1y_2| \geq 0,$$

and then (6.1) becomes

(6.2)
$$E_1 E_2 = F^2 + (ac - b^2)(x_1y_2 - x_2y_1)^2.$$

We also have

(6.3)
$$f(x_1 - x_2, y_1 - y_2) = E_1 + E_2 \pm 2F.$$

Hence, using (6.2) and (6.3), we obtain

$$(f(x_1, y_1)f(x_2, y_2))^{1/2} f(x_1 - x_2, y_1 - y_2)$$
$$\geq (E_1 E_2)^{1/2}(E_1 + E_2 - 2F)$$
$$\geq (E_1 E_2)^{1/2}(2(E_1 E_2)^{1/2} - 2F)$$
$$= 2(E_1 E_2) - 2(E_1 E_2)^{1/2}F$$
$$= 2F^2 + 2(ac - b^2)(x_1y_2 - x_2y_1)^2$$
$$\qquad -2F(F^2 + (ac - b^2)(x_1y_2 - x_2y_1)^2)^{1/2}$$
$$= 2F^2 + 2(ac - b^2)(x_1y_2 - x_2y_1)^2$$
$$\qquad -2F^2 \left(1 + \frac{(ac - b^2)(x_1y_2 - x_2y_1)^2}{F^2}\right)^{1/2}$$

$$\begin{aligned} \geq \quad & 2F^2 + 2(ac - b^2)(x_1y_2 - x_2y_1)^2 \\ & -2F^2\left(1 + \frac{(ac - b^2)(x_1y_2 - x_2y_1)^2}{2F^2}\right) \\ = \quad & (ac - b^2)(x_1y_2 - x_2y_1)^2 \ . \end{aligned}$$

This completes the proof of (6.0).

7.   Let R, S, T be three real numbers, not all the same. Give a condition which is satisfied by one and only one of the three triples

(7.0)
$$\begin{cases} (R, S, T), \\ (T, -S + 2T, R - S + T), \\ (R - S + T, 2R - S, R). \end{cases}$$

Solution: We let (a, b, c) denote any one of the triples in (7.0) and show that exactly one of the three triples satisfies

(7.1) (i) $a \leq b < c$ or (ii) $a \geq b > c$.

We consider six cases.
Case (i): $R \leq S < T$.   Here $(a, b, c) = (R, S, T)$ satisfies (7.1)(i) but not (7.1)(ii), while the other two triples satisfy neither (7.1)(i) nor (ii) as

$$T < -S + 2T, \quad -S + 2T > R - S + T$$

and

$$R - S + T > 2R - S, \quad 2R - S < R \ .$$

Case (ii): $R < T \leq S$.   Here $(a, b, c) = (T, -S + 2T, R - S + T)$ satisfies (7.1)(ii) but not (7.1)(i), while the other two triples satisfy neither (7.1)(i) nor (ii) as

$$R < S, \quad S > T$$

and

$$R - S + T > 2R - S, \quad 2R - S < R \ .$$

Case (iii): $S < R \leq T$. Here $(a, b, c) = (R - S + T, 2R - S, R)$ satisfies (7.1)(ii) but not (7.1)(i), while the other two triples satisfy neither (7.1)(i) nor (ii) as

$$R > S, \quad S < T$$

and

$$T < -S + 2T, \quad -S + 2T > R - S + T .$$

Case (iv): $S \leq T < R$. Here $(a, b, c) = (T, -S + 2T, R - S + T)$ satisfies (7.1)(i) but not (7.1)(ii), while the other two triples satisfy neither (7.1)(i) nor (ii) as

$$R > S, \quad S \leq T$$

and

$$R - S + T < 2R - S, \quad 2R - S > R .$$

Case (v): $T \leq R < S$. Here $(a, b, c) = (R - S + T, 2R - S, R)$ satisfies (7.1)(i) but not (7.1)(ii), while the other two triples satisfy neither (7.1)(i) nor (ii) as

$$R < S, \quad S > T$$

and

$$T > -S + 2T, \quad -S + 2T < R - S + T .$$

Case(vi): $T < S \leq R$. Here $(a, b, c) = (R, S, T)$ satisfies (7.1)(ii) but not (7.1)(i), while the other two triples satisfy neither (7.1)(i) nor (ii) as

$$T > -S + 2T, \quad -S + 2T < R - S + T$$

and

$$R - S + T < 2R - S, \quad 2R - S \geq R .$$

8. Let $ax^2 + bxy + cy^2$ and $Ax^2 + Bxy + Cy^2$ be two positive-definite quadratic forms, which are not proportional. Prove that the form

(8.0) $$(aB - bA)x^2 + 2(aC - cA)xy + (bC - cB)y^2$$

is indefinite.

Solution: As $ax^2 + bxy + cy^2$ and $Ax^2 + Bxy + Cy^2$ are positive-definite we have

$$a > 0, \quad c > 0, \quad b^2 - 4ac < 0,$$
$$A > 0, \quad C > 0, \quad B^2 - 4AC < 0.$$

To show that the form

$$(aB - bA)x^2 + 2(aC - cA)xy + (bC - cB)y^2$$

is indefinite we must show that its discriminant

$$D = 4(aC - cA)^2 - 4(aB - bA)(bC - cB)$$

is positive. We first show that $D \geq 0$. This follows as

$$a^2 D = (2a(aC - cA) - b(aB - bA))^2 - (b^2 - 4ac)(aB - bA)^2 .$$

Moreover, $D > 0$ unless

$$aB - bA = aC - cA = 0$$

in which case

$$\frac{a}{A} = \frac{b}{B} = \frac{c}{C} .$$

This does not occur as $ax^2 + bxy + cy^2$ and $Ax^2 + Bxy + Cy^2$ are not proportional.

9.   Evaluate the limit

(9.0)
$$L = \lim_{n \to \infty} \frac{n}{2^n} \sum_{k=1}^{n} \frac{2^k}{k} .$$

Solution: We show that $L = 2$. For $n \geq 3$ we have

$$\frac{n}{2^n} \sum_{k=1}^{n} \frac{2^k}{k} = \frac{n}{2^n} \sum_{k=0}^{n-1} \frac{2^{n-k}}{n-k}$$

$$= \sum_{k=0}^{n-1} \frac{1}{2^k} \frac{n}{n-k}$$

$$= \sum_{k=0}^{n-1} \frac{1}{2^k} \left(1 + \frac{k}{n-k} \right)$$

$$= \sum_{k=0}^{n-1} \frac{1}{2^k} + \sum_{k=1}^{n-1} \frac{k}{2^k(n-k)} \ ,$$

and so

$$\left| \frac{n}{2^n} \sum_{k=1}^{n} \frac{2^k}{k} - \sum_{k=0}^{n-1} \frac{1}{2^k} \right| = \left| \sum_{k=1}^{n-1} \frac{k}{2^k(n-k)} \right|$$

$$= \sum_{k=1}^{n-1} \frac{k}{2^k(n-k)}$$

$$\leq \frac{1}{2(n-1)} + \sum_{k=2}^{n-1} \frac{k}{(k^2-k)(n-k)}$$

$$= \frac{1}{2(n-1)} + \sum_{k=2}^{n-1} \frac{1}{(k-1)(n-k)}$$

$$= \frac{1}{2(n-1)} + \frac{1}{n-1} \sum_{k=2}^{n-1} \left(\frac{1}{k-1} + \frac{1}{n-k} \right)$$

$$= \frac{1}{2(n-1)} + \frac{2}{n-1} \sum_{r=1}^{n-2} \frac{1}{r}$$

$$\leq \frac{1}{2(n-1)} + \frac{2}{n-1} \ln n \ .$$

As $n \to +\infty$, $\frac{1}{2(n-1)} + \frac{2}{(n-1)} \ln n \to 0$ and so

$$L = \lim_{n \to \infty} \frac{n}{2^n} \sum_{k=1}^{n} \frac{2^k}{k} = \lim_{n \to \infty} \sum_{k=0}^{n-1} \frac{1}{2^k} = \sum_{k=0}^{\infty} \frac{1}{2^k} = 2 \ .$$

10.   Prove that there does not exist a constant $c \geq 1$ such that

(10.0) $n^c \phi(n) \geq m^c \phi(m),$

for all positive integers n and m satisfying $n \geq m$.

Solution: Suppose there exists a constant $c \geq 1$ such that (10.0) holds for all positive integers m and n satisfying $n \geq m$. Let p be a prime with $p > 4c$. Then, we have

$$\frac{3}{4} \geq \frac{p+1}{2(p-1)} \qquad (\text{as } p > 4c \geq 4)$$

$$\geq \frac{\phi(p+1)}{\phi(p)} \qquad (\text{as } \phi(p+1) \leq (p+1)/2, \ \phi(p) = p-1)$$

$$\geq \left(\frac{p}{p+1}\right)^c \qquad (\text{by } (10.0))$$

$$= \left(1 - \frac{1}{p+1}\right)^c$$

$$\geq 1 - \frac{c}{p+1} \qquad (\text{using } x^c - 1 \geq c(x-1), \ x > 0)$$

$$> 1 - \frac{c}{p}$$

$$> \frac{3}{4} \qquad (\text{as } p > 4c),$$

which is impossible, and no such c exists.

11.   Let D be a squarefree integer greater than 1 for which there exist positive integers A_1, A_2, B_1, B_2 such that

(11.0) $\begin{cases} D = & A_1^2 + B_1^2 &= A_2^2 + B_2^2, \\ & (A_1, B_1) & \neq (A_2, B_2). \end{cases}$

Prove that neither

$$2D(D + A_1 A_2 + B_1 B_2)$$

nor

$$2D(D + A_1 A_2 - B_1 B_2)$$

is the square of an integer.

Solution: Suppose that $2D(D + A_1 A_2 + \epsilon B_1 B_2) = X^2$, where X is an integer and $\epsilon = \pm 1$. We consider two cases according as D is odd or even.

If D is odd, as it is squarefree, $2D$ divides X, say $X = 2DU$, where U is an integer, and so

$$(11.1) \qquad D + A_1 A_2 + \epsilon B_1 B_2 = 2DU^2 .$$

Next we have

$$
\begin{aligned}
2D(D - A_1 A_2 - \epsilon B_1 B_2) &= 2D \frac{(D^2 - (A_1 A_2 + \epsilon B_1 B_2)^2)}{D + A_1 A_2 + \epsilon B_1 B_2} \\
&= \frac{2D(A_1 B_2 - \epsilon A_2 B_1)^2}{D + A_1 A_2 + \epsilon B_1 B_2} \, ,
\end{aligned}
$$

that is

$$(11.2) \qquad 2D(D - A_1 A_2 - \epsilon B_1 B_2) = \left(\frac{A_1 B_2 - \epsilon A_2 B_1}{U} \right)^2 .$$

Since the left side of (11.2) is an integer and the right side is the square of a rational number, the right side of (11.2) must in fact be the square of an integer. Hence, there is an integer Z such that

$$(11.3) \qquad 2D(D - A_1 A_2 - \epsilon B_1 B_2) = Z^2 ,$$

$$(11.4) \qquad A_1 B_2 - \epsilon A_2 B_1 = UZ .$$

From (11.3), as above, we see that $2D$ divides Z, so there exists V such that $Z = 2DV$. Then (11.3) and (11.4) become

$$(11.5) \qquad D - A_1 A_2 - \epsilon B_1 B_2 = 2DV^2 ,$$

(11.6) $$A_1 B_2 - \epsilon A_2 B_1 = 2DUV .$$

Adding (11.1) and (11.5) we obtain $2D = 2DU^2 + 2DV^2$, so that $U^2 + V^2 = 1$, giving

(11.7) $$(U, V) = (\pm 1, 0) \quad \text{or} \quad (0, \pm 1) .$$

Now from (11.1), (11.5) and (11.6), we have

$$\begin{cases} A_1 A_2 + \epsilon B_1 B_2 &= D(U^2 - V^2) , \\ -\epsilon B_1 A_2 + A_1 B_2 &= 2DUV . \end{cases}$$

Solving these equations for A_2 and B_2 gives

(11.8) $$A_2 = (U^2 - V^2)A_1 - 2\epsilon UV B_1, \quad B_2 = 2UV A_1 + \epsilon(U^2 - V^2)B_1 .$$

Using the values for (U, V) given in (11.7), we obtain from (11.8) $(A_2, B_2) = \pm(A_1, \epsilon B_1)$, which is clearly impossible as A_1, A_2, B_1, B_2 are positive and $(A_1, B_1) \neq (A_2, B_2)$.

The case when D is even can be treated similarly.

12.   Let **Q** and **R** denote the fields of rational and real numbers respectively. Let **K** and **L** be the smallest subfields of **R** which contain both **Q** and the real numbers

$$\sqrt{1985 + 31\sqrt{1985}} \quad \text{and} \quad \sqrt{3970 + 64\sqrt{1985}} ,$$

respectively. Prove that **K** = **L**.

Solution: We set

(12.1) $$\begin{cases} \alpha_+ = \sqrt{1985 + 31\sqrt{1985}} \approx 58.018, \\ \alpha_- = \sqrt{1985 - 31\sqrt{1985}} \approx 24.573, \end{cases}$$

(12.2) $$\begin{cases} \beta_+ = \sqrt{3970 + 64\sqrt{1985}} \approx 82.591, \\ \beta_- = \sqrt{3970 - 64\sqrt{1985}} \approx 33.445 . \end{cases}$$

It is easy to check that

$$(12.3) \qquad \alpha_+\alpha_- = 32\sqrt{1985}, \quad \beta_+\beta_- = 62\sqrt{1985},$$

$$(12.4) \qquad \begin{cases} (\alpha_+ + \alpha_-)^2 = 3970 + 64\sqrt{1985}, \\ (\alpha_+ - \alpha_-)^2 = 3970 - 64\sqrt{1985}, \end{cases}$$

from which we obtain

$$(12.5) \qquad \alpha_+ + \alpha_- = \beta_+, \quad \alpha_+ - \alpha_- = \beta_-.$$

Writing $\mathbf{Q}(\gamma_1, \ldots, \gamma_n)$ for the smallest subfield of $\mathbf{R}$ containing both $\mathbf{Q}$ and the real numbers $\gamma_1, \ldots, \gamma_n$, we have

$$
\begin{aligned}
\mathbf{Q}(\alpha_+) &= \mathbf{Q}(\alpha_+, \alpha_+^2) \\
&= \mathbf{Q}(\alpha_+, \sqrt{1985}) && \text{(by (12.1))} \\
&= \mathbf{Q}(\alpha_+, \alpha_-) && \text{(by (12.3))} \\
&\supseteq \mathbf{Q}(\alpha_+ + \alpha_-) \\
&= \mathbf{Q}(\beta_+) && \text{(by (12.5))} \\
&= \mathbf{Q}(\beta_+, \beta_+^2) \\
&= \mathbf{Q}(\beta_+, \sqrt{1985}) && \text{(by (12.2))} \\
&= \mathbf{Q}(\beta_+, \beta_-) && \text{(by (12.3))} \\
&\supseteq \mathbf{Q}(\beta_+ + \beta_-) \\
&= \mathbf{Q}(\alpha_+), && \text{(by (12.5))}
\end{aligned}
$$

so that $\mathbf{K} = \mathbf{Q}(\alpha_+) = \mathbf{Q}(\beta_+) = \mathbf{L}$.

13. Let k and l be positive integers such that

$$GCD(k,5) = GCD(l,5) = GCD(k,l) = 1$$

and

$$-k^2 + 3kl - l^2 = F^2, \quad \text{where } GCD(F,5) = 1.$$

Prove that the pair of equations

$$(13.0) \qquad \begin{cases} k = x^2 + y^2, \\ l = x^2 + 2xy + 2y^2, \end{cases}$$

has exactly two solutions in integers x and y.

Solution: We have

$$F^2 \equiv 4k^2 + 8kl + 4l^2 \equiv 4(k+l)^2 \pmod 5$$

so that $F \equiv \pm 2(k+l) \pmod 5$. Replacing F by $-F$, if necessary, we may suppose that

(13.1) $$F \equiv 2(k+l) \pmod 5 .$$

Then we have

(13.2) $$\begin{cases} 4k - l - 2F & \equiv \ 0 \pmod 5, \\ -3k + 2l - F & \equiv \ 0 \pmod 5, \\ k + l + 2F & \equiv \ 0 \pmod 5, \end{cases}$$

and we may define integers R, S, T by

(13.3) $$\begin{cases} 5R & = \ 4k - l - 2F, \\ 5S & = \ -3k + 2l - F, \\ 5T & = \ k + l + 2F . \end{cases}$$

Further, we have

$$\begin{aligned} 25(RT - S^2) & = (4k - l - 2F)(k + l + 2F) - (-3k + 2l - F)^2 \\ & = -5k^2 + 15kl - 5l^2 - 5F^2 \\ & = 0 , \end{aligned}$$

so that

(13.4) $$RT = S^2 .$$

We now treat three cases:

$$\text{(i) } R = S = 0, \quad \text{(ii) } R \neq 0, S = 0, \quad \text{(iii) } S \neq 0 .$$

Case (i): $R = S = 0$. From (13.3) we have $4k - l - 2F = 0$, and $-3k + 2l - F = 0$, so that $k = F, l = 2F$. But k, l are positive coprime integers, so

$F = 1, k = 1, l = 2$. In this case (13.0) has two solutions $(x, y) = \pm(0, 1)$.

Case (ii): $R \neq 0, S = 0$. From (13.4) we have $T = 0$, and so from (13.3) we obtain

$$\begin{cases} -3k + 2l - F &= 0 \,, \\ k + l + 2F &= 0 \,, \end{cases}$$

so that $k = l = -F$. As k, l are positive coprime integers we have $F = -1, k = l = 1$. In this case (13.0) has two solutions $(x, y) = \pm(1, 0)$.

Case(iii): $S \neq 0$. From (13.4) we have $RT > 0$. If $R < 0$ then $T < 0$ and we have $k = R + T < 0$, contradicting $k \geq 1$. Hence R and T are positive integers. Next, observe that

$$(4k - l - 2F)(4k - l + 2F) = (4k - l)^2 - 4F^2 = 5(l - 2k)^2 \,,$$

so that

$$(13.5) \qquad\qquad R(4k - l + 2F) = (l - 2k)^2 \,.$$

Clearly, we have $4k - l + 2F \neq 0$, otherwise $5R = -4F$ and so $5 \mid F$, contradicting $GCD(F, 5) = 1$. Hence we may define nonnegative integers a, b, c by

$$(13.6) \qquad\qquad 2^a \parallel R, \quad 2^b \parallel 4k - l + 2F, \quad 2^c \parallel l - 2k \,.$$

We have from (13.5) and (13.6)

$$(13.7) \qquad\qquad a + b = 2c$$

and

$$(13.8) \qquad\qquad \frac{R}{2^a} \frac{4k - l + 2F}{2^b} = \left(\frac{l - 2k}{2^c}\right)^2 \,,$$

where

$$\frac{R}{2^a}, \quad \frac{4k - l + 2F}{2^b}, \quad \frac{|l - 2k|}{2^c}$$

are odd positive integers. Suppose that

$$GCD\left(\frac{R}{2^a}, \frac{4k - l + 2F}{2^b}\right) > 1 \,.$$

Then there is an odd prime p which divides $R/2^a$ and $(4k - l + 2F)/2^b$, and thus p divides $4k - l - 2F, 4k - l + 2F$, and $l - 2k$, giving successively

$$p \mid 8k - 2l, \quad p \mid 4k - l, \quad p \mid 2k, \quad p \mid k, \quad p \mid l \,,$$

contradicting $GCD(k,l) = 1$. Hence we have

$$(13.9) \qquad\qquad GCD\left(\frac{R}{2^a}, \frac{4k - l + 2F}{2^b}\right) = 1 \,.$$

From (13.8) and (13.9) we see that

$$(13.10) \qquad\qquad \frac{R}{2^a} = X^2 \,,$$

for some integer X. Next we show that a is even. This is clear if $a = 0$ so we may suppose that $a \geq 1$. Thus $2 \mid R$ and so l is even. As $GCD(k,l) = 1$ we have k odd. Then, taking $-k^2 + 3kl - l^2 = F^2$ successively modulo 2,4 and 8, we get

$$(13.11) \qquad\qquad F \equiv 1 \pmod 2 \,,$$

$$(13.12) \qquad\qquad l \equiv 2 \pmod 4 \,,$$

$$(13.13) \qquad\qquad l \equiv 2k \pmod 8 \,.$$

Thus we have $4k - l \pm 2F \equiv 0 \pmod 4$ and so $a \geq 2, b \geq 2$. Also we have

$$2^{\min(a,b)} \mid (4k - l + 2F) - (4k - l - 2F) = 4F \,,$$

and so as F is odd we have $\min(a,b) \leq 2$. If $a \leq b$ then we have $a \leq 2$, which implies that $a = 2$. If $b < a$ then $b \leq 2$, which implies that $b = 2$, $a = 2c - 2$. In both cases a is even as asserted.

Setting $a = 2d$, $x_0 = 2^d X$, we have $R = x_0^2$. Then from (13.4) we deduce that $T = y_0^2$, $S = \pm x_0 y_0$. Changing the sign of x_0 if necessary we may suppose that $S = x_0 y_0$. Thus we obtain $x_0^2 + y_0^2 = R + T = (5R + 5T)/5 = (4k - l - 2F + k + l + 2F)/5 = k$ and $x_0^2 + 2x_0 y_0 + 2y_0^2 = R + 2S + 2T = l$, so that (x_0, y_0) is a solution of (13.0).

Now let (x, y) be *any* solution of (13.0). Then using (13.0) we have

$$F^2 = -k^2 + 3kl - l^2 = (x^2 + xy - y^2)^2 ,$$

so that (with F chosen to satisfy (13.1))

(13.14)
$$x^2 + xy - y^2 = \pm F .$$

Solving (13.0) and (13.14) for x^2, xy, y^2, we get

(13.15)
$$\begin{cases} 5x^2 &=& 4k - l \pm 2F, \\ 5xy &=& -3k + 2l \pm F, \\ 5y^2 &=& k + l \mp 2F . \end{cases}$$

As

$$F \equiv 2(k + l) \not\equiv 0 \pmod 5$$

the lower signs must hold in (13.15), and so

(13.16)
$$\begin{cases} x^2 &=& (4k - l - 2F)/5 , \\ xy &=& (-3k + 2l - F)/5 , \\ y^2 &=& (k + l + 2F)/5 . \end{cases}$$

Since this is true for any solution of (13.0) we must have that (13.16) holds with x, y replaced by x_0, y_0 respectively. This means that

$$x^2 = x_0^2, \quad xy = x_0 y_0, \quad y^2 = y_0^2 ,$$

giving

$$(x, y) = (x_0, y_0), \quad \text{or} \quad (-x_0, -y_0) ,$$

and proving that (13.0) has exactly two integral solutions.

14. Let r and s be non-zero integers. Prove that the equation

(14.0)
$$(r^2 - s^2)x^2 - 4rsxy - (r^2 - s^2)y^2 = 1$$

has no solutions in integers x and y.

Solution: We suppose that x and y are integers satisfying (14.0). Factoring
the left side of (14.0), we obtain

(14.1) $((r - s)x - (r + s)y)((r + s)x + (r - s)y) = 1$.

As each factor on the left side of (14.1) is an integer, we see that

(14.2) $\begin{cases} (r - s)x - (r + s)y = \epsilon , \\ (r + s)x + (r - s)y = \epsilon , \end{cases}$

where $\epsilon = \pm 1$. Solving (14.2) for x and y, we obtain

(14.3) $x = \dfrac{r\epsilon}{r^2 + s^2} , \quad y = \dfrac{-s\epsilon}{r^2 + s^2} .$

Hence we have $(x^2 + y^2)(r^2 + s^2) = 1$, so that $r^2 + s^2 = 1$, that is

$$(r, s) = (\pm 1, 0) \quad \text{or} \quad (0, \pm 1) ,$$

which is impossible as r and s are both non-zero, thus showing that (14.0)
has no integral solutions.

15.   Evaluate the integral

(15.0) $$I = \int_0^1 \ln x \ln(1 - x)\, dx .$$

Solution: The function $\ln x \ln(1 - x)$ is continuous for $0 < x < 1$, but is not
defined at $x = 0$ and $x = 1$, so that

(15.1) $$I = \lim_{\substack{\epsilon \to 0^+ \\ \delta \to 0^+}} \int_\epsilon^{1-\delta} \ln x \ln(1 - x) dx .$$

For x satisfying

(15.2) $$0 < \epsilon \le x \le 1 - \delta < 1 ,$$

and n a positive integer, we have

$$- \ln(1 - x) = \sum_{k=1}^{\infty} \frac{x^k}{k} = \sum_{k=1}^{n} \frac{x^k}{k} + x^{n+1} \sum_{k=n+1}^{\infty} \frac{x^{k-(n+1)}}{k} \, ,$$

and so

$$\left| \ln(1 - x) + \sum_{k=1}^{n} \frac{x^k}{k} \right| = x^{n+1} \sum_{k=0}^{\infty} \frac{x^k}{n + 1 + k}$$

$$\leq \frac{x^{n+1}}{n + 1} \sum_{k=0}^{\infty} x^k$$

$$= \frac{x^{n+1}}{(n + 1)(1 - x)} \, .$$

Thus we have

$$(15.3) \quad \left| \int_{\epsilon}^{1-\delta} \ln x \left(\ln(1 - x) + \sum_{k=1}^{n} \frac{x^k}{k} \right) dx \right|$$

$$\leq \frac{1}{(n + 1)} \int_{\epsilon}^{1-\delta} (-\ln x) \frac{x^{n+1}}{(1 - x)} dx \, .$$

Now, for $y \geq 1$, we have

$$(15.4) \quad 0 \leq \ln y = \int_{1}^{y} \frac{dt}{t} \leq \int_{1}^{y} dt = y - 1 \, .$$

Taking $y = 1/x$ in (15.4), we have

$$(15.5) \quad 0 \leq - \ln x = \ln \left(\frac{1}{x} \right) \leq \frac{1}{x} - 1 = \frac{1 - x}{x} \, .$$

Using the inequality (15.5) in (15.3) we deduce

$$\left| \int_{\epsilon}^{1-\delta} \ln x \ln(1 - x) dx + \sum_{k=1}^{n} \frac{1}{k} \int_{\epsilon}^{1-\delta} x^k \ln x \, dx \right|$$

$$\leq \frac{1}{n + 1} \int_{\epsilon}^{1-\delta} x^n dx$$

$$< \frac{1}{n + 1} \int_{0}^{1} x^n dx$$

$$= \frac{1}{(n + 1)^2} \, ,$$

and letting $n \to \infty$, we obtain

(15.6) $$\int_\epsilon^{1-\delta} \ln x \ln(1-x) dx = -\sum_{k=1}^\infty \frac{1}{k} \int_\epsilon^{1-\delta} x^k \ln x \, dx \, .$$

As

$$\frac{d}{dx}\left(\frac{x^{k+1}\ln x}{k+1} - \frac{x^{k+1}}{(k+1)^2}\right) = x^k \ln x \, ,$$

by the fundamental theorem of calculus, we have

$$\int_\epsilon^{1-\delta} x^k \ln x \, dx = \frac{(1-\delta)^{k+1}\ln(1-\delta)}{k+1} - \frac{(1-\delta)^{k+1}}{(k+1)^2}$$
$$- \frac{\epsilon^{k+1}\ln \epsilon}{k+1} + \frac{\epsilon^{k+1}}{(k+1)^2} \, ,$$

so that by (15.6)

$$\int_\epsilon^{1-\delta} \ln x \ln(1-x) dx = \ln \epsilon \sum_{k=1}^\infty \frac{\epsilon^{k+1}}{k(k+1)} - \sum_{k=1}^\infty \frac{\epsilon^{k+1}}{k(k+1)^2}$$
$$- \ln(1-\delta) \sum_{k=1}^\infty \frac{(1-\delta)^{k+1}}{k(k+1)} + \sum_{k=1}^\infty \frac{(1-\delta)^{k+1}}{k(k+1)^2} \, ,$$

that is

(15.7) $$\int_\epsilon^{1-\delta} \ln x \ln(1-x) dx$$
$$= (\ln \epsilon)A(\epsilon) - B(\epsilon) - (\ln(1-\delta))A(1-\delta) + B(1-\delta) \, ,$$

where, for $0 < y < 1$, $A(y)$ and $B(y)$ are defined by

(15.8) $$A(y) = \sum_{k=1}^\infty \frac{y^{k+1}}{k(k+1)} \, ,$$

(15.9) $$B(y) = \sum_{k=1}^\infty \frac{y^{k+1}}{k(k+1)^2} \, ,$$

We next show that

(15.10)
$$\lim_{\epsilon \to 0+} (\ln \epsilon) A(\epsilon) = 0 \ ,$$

(15.11)
$$\lim_{\delta \to 0+} (\ln(1 - \delta)) A(1 - \delta) = 0 \ ,$$

(15.12)
$$\lim_{\epsilon \to 0+} B(\epsilon) = 0 \ ,$$

(15.13)
$$\lim_{\delta \to 0+} B(1 - \delta) = 2 - \frac{\pi^2}{6} \ ,$$

so that (15.1) and (15.7) give

(15.14)
$$I = 2 - \frac{\pi^2}{6} \ ,$$

as asserted in the HINTS.

Before proving (15.10)-(15.13) we show that

(15.15)
$$\lim_{\epsilon \to 0+} (\ln \epsilon) \ln(1 - \epsilon) = 0 \ .$$

For $0 < \epsilon < 1$ we have

$$-\ln(1 - \epsilon) = \epsilon + \frac{\epsilon^2}{2} + \frac{\epsilon^3}{3} + \cdots \begin{cases} > \epsilon \ , \\ < \ \epsilon + \epsilon^2 + \epsilon^3 + \cdots \end{cases} = \frac{\epsilon}{1 - \epsilon} \ ,$$

so that

$$-\epsilon \ln \epsilon < (\ln \epsilon) \ln(1 - \epsilon) < -\frac{\epsilon \ln \epsilon}{1 - \epsilon} \ ,$$

from which (15.15) follows, as

(15.16)
$$\lim_{\epsilon \to 0+} \epsilon \ln \epsilon = 0 \ .$$

Now for $0 < \epsilon < 1$ we have

$$
\begin{aligned}
A(\epsilon) &= \epsilon \sum_{k=1}^{\infty} \frac{\epsilon^k}{k} - \sum_{k=1}^{\infty} \frac{\epsilon^{k+1}}{k+1} \\
&= -\epsilon \ln(1 - \epsilon) + \ln(1 - \epsilon) + \epsilon \\
&= (1 - \epsilon) \ln(1 - \epsilon) + \epsilon ,
\end{aligned}
$$

so that

$$
\lim_{\epsilon \to 0^+} (\ln \epsilon) A(\epsilon) = 0 .
$$

This proves (15.10).

Next we have, by Abel's theorem,

$$
\lim_{\delta \to 0^+} A(1 - \delta) = \lim_{\delta \to 0^+} \sum_{k=1}^{\infty} \frac{(1 - \delta)^{k+1}}{k(k+1)} = \sum_{k=1}^{\infty} \frac{1}{k(k+1)} = 1 ,
$$

so that

$$
\lim_{\delta \to 0^+} (\ln(1 - \delta)) A(1 - \delta) = \ln 1 = 0 .
$$

This proves (15.11). Also we have

$$
|B(\epsilon)| \le \epsilon \sum_{k=1}^{\infty} \frac{1}{k(k+1)^2}
$$

so that

$$
\lim_{\epsilon \to 0^+} B(\epsilon) = 0 ,
$$

proving (15.12). Finally, by Abel's theorem, we have

$$
\begin{aligned}
\lim_{\delta \to 0^+} B(1 - \delta) &= \lim_{\delta \to 0^+} \sum_{k=1}^{\infty} \frac{(1 - \delta)^{k+1}}{k(k+1)^2} \\
&= \sum_{k=1}^{\infty} \frac{1}{k(k+1)^2} \\
&= \sum_{k=1}^{\infty} \left(\frac{1}{k(k+1)} - \frac{1}{(k+1)^2} \right) \\
&= 1 - \left(\frac{\pi^2}{6} - 1 \right) \\
&= 2 - \frac{\pi^2}{6} ,
\end{aligned}
$$

proving (15.13), and completing the proof of (15.14).

16. Solve the recurrence relation

(16.0)
$$\sum_{k=1}^{n} \binom{n}{k} a(k) = \frac{n}{n+1}, \quad n = 1, 2, \dots .$$

Solution: We make the inductive hypothesis that $a(n) = (-1)^{n+1}/(n+1)$ for all positive integers n satisfying $1 \le n \le m$. This hypothesis is true for $m = 1$ as $a(1) = 1/2$. Now, by (16.0) and the inductive hypothesis, we have

$$a(m+1) = \frac{m+1}{m+2} - \sum_{k=1}^{m} \binom{m+1}{k} \frac{(-1)^{k+1}}{k+1} .$$

Thus we must show that

$$\sum_{k=1}^{m} \binom{m+1}{k} \frac{(-1)^{k+1}}{k+1} = \frac{m+1-(-1)^m}{m+2} ,$$

or equivalently

$$\sum_{k=1}^{m+1} \binom{m+1}{k} \frac{(-1)^{k+1}}{k+1} = \frac{m+1}{m+2} .$$

By the binomial theorem, we have for any real number x

(16.1)
$$(1+x)^{m+1} = \sum_{k=0}^{m+1} \binom{m+1}{k} x^k .$$

Integrating (16.1) with respect to x, we obtain

(16.2)
$$\frac{(1+x)^{m+2}}{m+2} = \sum_{k=0}^{m+1} \binom{m+1}{k} \frac{x^{k+1}}{k+1} + \frac{1}{m+2} .$$

Taking $x = -1$ in (16.2) we have

$$\sum_{k=0}^{m+1} \binom{m+1}{k} \frac{(-1)^{k+1}}{k+1} = -\frac{1}{m+2} ,$$

and so

$$\sum_{k=1}^{m+1} \binom{m+1}{k} \frac{(-1)^{k+1}}{k+1} = 1 - \frac{1}{m+2} = \frac{m+1}{m+2}$$

as required. The result now follows by the principle of mathematical induction.

17. Let n and k be positive integers. Let p be a prime such that

$$p > (n^2 + n + k)^2 + k \ .$$

Prove that the sequence

(17.0) $$n^2, \ n^2 + 1, \ n^2 + 2, \ldots, n^2 + l \ ,$$

where $l = (n^2 + n + k)^2 - n^2 + k$, contains a pair of integers $(m, m+k)$ such that

$$\left(\frac{m}{p}\right) = \left(\frac{m+k}{p}\right) = 1 \ .$$

Solution: As n and k are positive integers and $p > (n^2 + n + k)^2 + k$, none of the integers of the sequence (17.0) is divisible by p. If $\left(\frac{n^2+k}{p}\right) = 1$ we can take $(m, m+k) = (n^2, n^2 + k)$. If $\left(\frac{(n+1)^2+k}{p}\right) = 1$ we can take $(m, m+k) = ((n+1)^2, (n+1)^2 + k)$. Finally, if

$$\left(\frac{n^2 + k}{p}\right) = \left(\frac{(n+1)^2 + k}{p}\right) = -1 \ ,$$

we can take $(m, m+k) = ((n^2 + n + k)^2, (n^2 + n + k)^2 + k)$, as

$$\left(\frac{(n^2 + n + k)^2 + k}{p}\right) = \left(\frac{(n^2 + k)((n+1)^2 + k)}{p}\right)$$

$$= \left(\frac{n^2 + k}{p}\right)\left(\frac{(n+1)^2 + k}{p}\right)$$

$$= (-1)(-1) = 1 \ .$$

This establishes the existence of a pair of integers as required.

18. Let

$$a_n = \frac{1}{4n+1} + \frac{1}{4n+3} - \frac{1}{2n+2}, \quad n = 0, 1, \ldots .$$

Does the infinite series $\sum_{n=0}^{\infty} a_n$ converge, and if so, what is its sum?

Solution: Let $s(N) = \sum_{n=0}^{N} a_n$, $N = 0, 1, \ldots$. We have

$$
\begin{aligned}
s(N) &= \sum_{n=0}^{N} \left(\frac{1}{4n+1} + \frac{1}{4n+3} - \frac{1}{2n+2} \right) \\
&= \sum_{n=0}^{N} \left(\frac{1}{4n+1} - \frac{1}{4n+2} + \frac{1}{4n+3} - \frac{1}{4n+4} + \frac{1}{4n+2} - \frac{1}{4n+4} \right) \\
&= \sum_{m=1}^{4N+4} \frac{(-1)^{m-1}}{m} + \frac{1}{2} \sum_{m=1}^{2N+2} \frac{(-1)^{m-1}}{m} .
\end{aligned}
$$

Letting $N \to \infty$ we have

$$
\begin{aligned}
\lim_{N \to \infty} s(N) &= \sum_{m=1}^{\infty} \frac{(-1)^{m-1}}{m} + \frac{1}{2} \sum_{m=1}^{\infty} \frac{(-1)^{m-1}}{m} \\
&= \frac{3}{2} \sum_{m=1}^{\infty} \frac{(-1)^{m-1}}{m} \\
&= \frac{3}{2} \ln 2 ,
\end{aligned}
$$

so that $\sum_{n=0}^{\infty} a_n$ converges with sum $\frac{3}{2} \ln 2$.

19. Let $a_1, \ldots, a_m$ be $m \ (\geq 2)$ real numbers. Set

$$A_n = a_1 + a_2 + \ldots + a_n, \quad n = 1, 2, \ldots, m .$$

Prove that

$$(19.0) \qquad \sum_{n=2}^{m} \left(\frac{A_n}{n} \right)^2 \le 12 \sum_{n=1}^{m} a_n^2 .$$

Solution: For $n = 1, 2, \ldots, m$ we have

$$
\begin{aligned}
\left(\frac{A_n}{n} \right)^2 &= \left(a_n + \frac{A_n}{n} - a_n \right)^2 \\
&\le 2a_n^2 + 2 \left(\frac{A_n}{n} - a_n \right)^2 \\
&= 4a_n^2 + 2 \left(\frac{A_n}{n} \right)^2 - 4a_n \frac{A_n}{n} ,
\end{aligned}
$$

and so

$$(19.1) \qquad \sum_{n=1}^{m} \left(\frac{A_n}{n} \right)^2 \le 4 \sum_{n=1}^{m} a_n^2 + 2 \sum_{n=1}^{m} \left(\frac{A_n}{n} \right)^2 - 4 \sum_{n=1}^{m} a_n \frac{A_n}{n} .$$

But as

$$-2a_n A_n = -(A_n^2 - A_{n-1}^2) - a_n^2 \le -(A_n^2 - A_{n-1}^2)$$

we have

$$
\begin{aligned}
-2 \sum_{n=1}^{m} a_n \frac{A_n}{n} &\le - \sum_{n=1}^{m} \frac{(A_n^2 - A_{n-1}^2)}{n} \\
&= - \sum_{n=1}^{m-1} \frac{A_n^2}{n(n+1)} - \frac{A_m^2}{m}
\end{aligned}
$$

that is

$$(19.2) \qquad -2 \sum_{n=1}^{m} a_n \frac{A_n}{n} \le - \sum_{n=1}^{m} \frac{A_n^2}{n(n+1)} .$$

Using (19.2) in (19.1) we obtain

$$\sum_{n=1}^{m} \left(\frac{A_n}{n} \right)^2 \leq 4 \sum_{n=1}^{m} a_n^2 + 2 \sum_{n=1}^{m} \left(\frac{A_n}{n} \right)^2 - 2 \sum_{n=1}^{m} \frac{A_n^2}{n(n+1)}$$

$$= 4 \sum_{n=1}^{m} a_n^2 + 2 \sum_{n=1}^{m} \frac{A_n^2}{n^2(n+1)},$$

that is

$$(19.3) \qquad \sum_{n=1}^{m} \left(1 - \frac{2}{n+1} \right) \left(\frac{A_n}{n} \right)^2 \leq 4 \sum_{n=1}^{m} a_n^2.$$

The inequality (19.0) now follows from (19.3) by noting that $1 - \frac{2}{n+1} = 0$ when $n = 1$, and $1 - \frac{2}{n+1} \geq \frac{1}{3}$ for $n \geq 2$.

20.   Evaluate the sum

$$S = \sum_{k=0}^{n} \frac{\binom{n}{k}}{\binom{2n-1}{k}}$$

for all positive integers n.

Solution: We have

$$\frac{\binom{n}{k}}{\binom{2n}{k}} - \frac{\binom{n}{k+1}}{\binom{2n}{k+1}} = \frac{n!\,(2n-k)!}{(n-k)!\,2n!} - \frac{n!\,(2n-k-1)!}{(n-k-1)!\,2n!}$$

$$= \frac{n!\,(2n-1-k)!}{(n-k)!\,(2n-1)!} \left(\frac{(2n-k)}{2n} - \frac{(n-k)}{2n} \right)$$

$$= \frac{1}{2} \frac{\binom{n}{k}}{\binom{2n-1}{k}},$$

so that

$$S = 2 \sum_{k=0}^{n} \left(\frac{\binom{n}{k}}{\binom{2n}{k}} - \frac{\binom{n}{k+1}}{\binom{2n}{k+1}} \right)$$

$$= 2 \left(\frac{\binom{n}{0}}{\binom{2n}{0}} - \frac{\binom{n}{n+1}}{\binom{2n}{n+1}} \right)$$

$$= 2 .$$

21.   Let a and b be coprime positive integers. For k a positive integer, let $N(k)$ denote the number of integral solutions to the equation

(21.0) $ax + by = k, \quad x \geq 0, \quad y \geq 0 .$

Evaluate the limit

$$L = \lim_{k \to +\infty} \frac{N(k)}{k} .$$

Solution: As a and b are coprime there are integers g and h such that

(21.1) $ag + bh = k .$

Then all solutions of $ax + by = k$ are given by

(21.2) $x = g + bt, \quad y = h - at, \quad t = 0, \pm 1, \pm 2, \ldots .$

Thus the solutions of (21.0) are given by (21.2) for those integral values of t satisfying

(21.3) $\frac{h}{a} \geq t \geq -\frac{g}{b} .$

Set

(21.4) $\lambda(b, g) = \begin{cases} 0, & \text{if } b \text{ divides } g , \\ 1, & \text{if } b \text{ does not divide } g , \end{cases}$

Then there are

$$\left[\frac{h}{a} \right] - \left[\frac{-g}{b} \right] - \lambda(b, g) + 1$$

values of t satisfying (21.3). Hence we have

$$(21.5) \qquad N(k) = \left[\frac{h}{a}\right] - \left[\frac{-g}{b}\right] - \lambda(b,g) + 1 \,,$$

and so

$$\left| N(k) - \frac{h}{a} - \frac{g}{b} \right| \le 1 + 1 + 1 + 1 = 4 \,,$$

giving, by (21.1),

$$(21.6) \qquad \left| \frac{N(k)}{k} - \frac{1}{ab} \right| \le \frac{4}{k} \,.$$

Letting $k \to +\infty$ in (21.6), we obtain $L = 1/ab$.

22. Let a, d and r be positive integers. For $k = 0, 1, \ldots$ set

$$(22.0) \qquad u_k = u_k(a,d,r) = \frac{1}{(a+kd)(a+(k+1)d)\ldots(a+(k+r)d)} \,.$$

Evaluate the sum

$$S = \sum_{k=0}^{n} u_k \,,$$

where n is a positive integer.

Solution: For $k = -1, 0, 1, \ldots$ we set

$$(22.1) \qquad v_k = v_k(a,d,r) = \frac{1}{(a+(k+1)d)\cdots(a+(k+r)d)rd} \,,$$

so that

$$v_k - v_{k+1}$$
$$= \frac{1}{(a+(k+2)d)\cdots(a+(k+r)d)rd} \left(\frac{1}{(a+(k+1)d)} - \frac{1}{(a+(k+r+1)d)} \right)$$
$$= \frac{1}{(a+(k+1)d)(a+(k+2)d)\cdots(a+(k+r)d)(a+(k+r+1)d)} \,,$$

that is $v_k - v_{k+1} = u_{k+1}$. Hence we have

$$S = \sum_{k=0}^{n} u_k = \sum_{k=-1}^{n-1} u_{k+1} = \sum_{k=-1}^{n-1} (v_k - v_{k+1}) = v_{-1} - v_n \, ,$$

that is

$$S = \frac{1}{rd} \left(\frac{1}{a(a+d)\cdots(a+(r-1)d)} - \frac{1}{(a+(n+1)d)\cdots(a+(n+r)d)} \right).$$

23.   Let $x_1, \ldots, x_n$ be n (> 1) real numbers. Set

$$x_{ij} = x_i - x_j \quad (1 \le i < j \le n).$$

Let F be a real-valued function of the $n(n-1)/2$ variables x_{ij} such that the inequality

(23.0) $$F(x_{11}, x_{12}, \ldots, x_{n-1\,n}) \le \sum_{k=1}^{n} x_k^2$$

holds for all $x_1, \ldots, x_n$.

Prove that equality cannot hold in (23.0) if $\sum_{k=1}^{n} x_k \ne 0$.

Solution: Set $M = (x_1 + \cdots + x_n)/n$, and replace each x_i by $x_i - M$ in (23.0). Then (23.0) gives the stronger inequality

$$F(x_{11}, x_{12}, \ldots, x_{n-1\,n}) \le \sum_{k=1}^{n} (x_k - M)^2 = \sum_{k=1}^{n} x_k^2 - \frac{1}{n} \left(\sum_{k=1}^{n} x_k \right)^2.$$

Hence if $x_1, \ldots, x_n$ are chosen so that $\sum_{k=1}^{n} x_k \ne 0$, equality cannot hold in (23.0).

24.   Let $a_1, \ldots, a_m$ be m (≥ 1) real numbers which are such that $\sum_{n=1}^{m} a_n \ne 0$. Prove the inequality

(24.0) $$\left(\sum_{n=1}^{m} n a_n^2 \right) \Big/ \left(\sum_{n=1}^{m} a_n \right)^2 > \frac{1}{2\sqrt{m}} \, .$$

Solution: By the Cauchy-Schwarz inequality we have

$$(24.1) \qquad \left(\sum_{n=1}^{m} a_n \right)^2 = \left(\sum_{n=1}^{m} a_n \sqrt{n} \, \frac{1}{\sqrt{n}} \right)^2 \leq \sum_{n=1}^{m} n a_n^2 \, \sum_{n=1}^{m} \frac{1}{n} \, .$$

Next, we have

$$\sum_{n=1}^{m} \frac{1}{n} \leq 1 + \int_1^m \frac{dx}{x} \leq 1 + \int_1^m \frac{dx}{\sqrt{x}}$$
$$= 1 + (2\sqrt{m} - 2) = 2\sqrt{m} - 1 < 2\sqrt{m} \, .$$

We obtain (24.0) by using the latter inequality in (24.1).

25. Prove that there exist infinitely many positive integers which are not expressible in the form $n^2 + p$, where n is a positive integer and p is a prime.

Solution: We show that the integers $(3m + 2)^2$, $m = 1, 2, \ldots$, cannot be expressed in the form $n^2 + p$, where $n \geq 1$ and p is a prime. For suppose that

$$(3m + 2)^2 = n^2 + p \, ,$$

where $n \geq 1$ and p is a prime, then

$$(25.1) \qquad p = (3m + 2 - n)(3m + 2 + n) \, .$$

Since p is a prime and $0 < 3m + 2 - n < 3m + 2 + n$, we must have

$$(25.2) \qquad 3m + 2 - n = 1, \quad 3m + 2 + n = p \, .$$

Solving (25.2) for m and n we get

$$m = (p - 3)/6, \quad n = (p - 1)/2 \, ,$$

so that $p = 3(2m + 1)$. As p is prime, we must have $m = 0$, which contradicts $m \geq 1$.

26.   Evaluate the infinite series

$$S = \sum_{n=1}^{\infty} \arctan\left(\frac{2}{n^2}\right).$$

Solution: For $n \geq 1$ we have

$$\arctan\left(\frac{1}{n}\right) - \arctan\left(\frac{1}{n+2}\right) = \arctan\left(\frac{\frac{1}{n} - \frac{1}{n+2}}{1 + \frac{1}{n(n+2)}}\right)$$

$$= \arctan\left(\frac{2}{(n+1)^2}\right),$$

so that for $N \geq 2$ we have

$$\sum_{n=2}^{N} \arctan\left(\frac{2}{n^2}\right) = \sum_{n=1}^{N-1} \arctan\left(\frac{2}{(n+1)^2}\right)$$

$$= \sum_{n=1}^{N-1}\left(\arctan\left(\frac{1}{n}\right) - \arctan\left(\frac{1}{n+2}\right)\right)$$

$$= \arctan(1) + \arctan\left(\frac{1}{2}\right) - \arctan\left(\frac{1}{N}\right)$$

$$- \arctan\left(\frac{1}{N+1}\right).$$

Letting $N \to \infty$ we get

$$\sum_{n=2}^{\infty} \arctan\left(\frac{2}{n^2}\right) = \arctan(1) + \arctan\left(\frac{1}{2}\right) = \frac{\pi}{4} + \arctan\left(\frac{1}{2}\right)$$

and so

$$S = \frac{\pi}{4} + \arctan(2) + \arctan\left(\frac{1}{2}\right) = \frac{3\pi}{4}.$$

27.   Let $p_1, \ldots, p_n$ denote n (≥ 1) distinct integers and let $f_n(x)$ be the polynomial of degree n given by

$$f_n(x) = (x - p_1)(x - p_2) \ldots (x - p_n).$$

Prove that the polynomial

$$g_n(x) = (f_n(x))^2 + 1$$

cannot be expressed as the product of two non-constant polynomials with integral coefficients.

Solution: Suppose that $g_n(x)$ can be expressed as the product of two non-constant polynomials with integral coefficients, say

$$(27.1) \qquad\qquad g_n(x) = h(x)k(x) \, .$$

Neither $h(x)$ nor $k(x)$ has a real root as $g_n(x) > 0$ for all real x. Thus, neither $h(x)$ nor $k(x)$ can change sign as x takes on all real values, and we may suppose that

$$(27.2) \qquad\qquad h(x) > 0, \quad k(x) > 0, \quad \text{for all real } x \, .$$

Since $g_n(p_i) = 1$, $i = 1, 2, \ldots, n$, we have $h(p_i) = k(p_i) = 1$, $i = 1, 2, \ldots, n$. If the degree of either $h(x)$ or $k(x)$ were less than n, then the polynomial would have to be identically 1, which is not the case as $h(x)$ and $k(x)$ are non-constant polynomials. Hence both $h(x)$ and $k(x)$ have degree n, and

$$(27.3) \qquad\qquad \begin{cases} h(x) = 1 + a(x - p_1) \cdots (x - p_n) \, , \\ k(x) = 1 + b(x - p_1) \cdots (x - p_n) \, , \end{cases}$$

for integers a and b. Thus we have

$$(27.4) \quad \begin{aligned} &(x - p_1)^2 (x - p_2)^2 \cdots (x - p_n)^2 + 1 \\ &= 1 + (a + b)(x - p_1) \cdots (x - p_n) + ab(x - p_1)^2 \cdots (x - p_n)^2 \, . \end{aligned}$$

Equating coefficients of x^{2n} and x^n in (27.4) we obtain

$$(27.5) \qquad\qquad \begin{cases} ab & = 1 \,, \\ a + b & = 0 \,. \end{cases}$$

Thus we have a contradiction as no integers satisfy (27.5).

28.   Two people, A and B, play a game in which the probability that A wins is p, the probability that B wins is q, and the probability of a draw is r. At the beginning, A has m dollars and B has n dollars. At the end of each game the winner takes a dollar from the loser. If A and B agree to play until one of them loses all his/her money, what is the probabilty of A winning all the money?

Solution: Let $p(k)$, $k = 0, 1, \ldots$, denote the probability that A wins when he/she has k dollars. Clearly, we have

$$(28.1) \qquad\qquad p(0) = 0, \quad p(m + n) = 1 \,.$$

We want to determine $p(m)$. Consider A's chances of winning when he/she has $k + 1$ dollars. If A wins the next game, A's probability of ultimately winning is $ap(k + 2)$. If A loses the next game however, A's probability of ultimately winning is $bp(k)$, while if the game is drawn, A's probability of ultimately winning is $cp(k + 1)$. Hence we have

$$p(k + 1) = ap(k + 2) + bp(k) + cp(k + 1) \,.$$

As $a + b + c = 1$ we deduce that

$$ap(k + 2) - (a + b)p(k + 1) + bp(k) = 0 \,.$$

Soving this difference equation, we obtain

$$p(k) = \begin{cases} A + Bk & , \text{ if } a = b \,, \\ A + B(b/a)^k & , \text{ if } a \neq b \,, \end{cases}$$

where A and B are constants to be determined. Using (28.1) we obtain

$$\begin{cases} A = \quad 0, \quad B = 1/(m+n) & \text{, if } a = b , \\ A = -B = 1/\left(1 - (b/a)^{m+n}\right) & \text{, if } a \neq b , \end{cases}$$

so that

$$p(m) = \begin{cases} m/(m+n) & \text{, if } a = b , \\ \left(1 - (b/a)^m\right)/\left(1 - (b/a)^{m+n}\right) & \text{, if } a \neq b . \end{cases}$$

29. Let $f(x)$ be a monic polynomial of degree $n \geq 1$ with complex coefficients. Let $x_1, \ldots, x_n$ denote the n complex roots of $f(x)$. The discriminant $D(f)$ of the polynomial $f(x)$ is the complex number

$$(29.0) \qquad\qquad D(f) = \prod_{1 \leq i < j \leq n} (x_i - x_j)^2.$$

Express the discriminant of $f(x^2)$ in terms of $D(f)$.

Solution: As $x_1, \ldots, x_n$ are the n roots of $f(x)$, the $2n$ roots of $f(x^2)$ are

$$y_1 = \sqrt{x_1},\ y_2 = \sqrt{x_2},\ \ldots ,\ y_n = \sqrt{x_n},\ y_{n+1} = -\sqrt{x_1},\ \ldots ,\ y_{2n} = -\sqrt{x_n} .$$

Hence, the discriminant of $f(x^2)$ is

$$
\begin{aligned}
\prod_{1 \leq i < j \leq 2n} (y_i - y_j)^2 &= \prod_{1 \leq i < j \leq n} (y_i - y_j)^2 \prod_{1 \leq i \leq n < j \leq 2n} (y_i - y_j)^2 \\
&\qquad\qquad \prod_{n < i < j \leq 2n} (y_i - y_j)^2 \\
&= \prod_{1 \leq i < j \leq n} (\sqrt{x_i} - \sqrt{x_j})^2 \prod_{1 \leq i \leq n < j \leq 2n} (\sqrt{x_i} + \sqrt{x_{j-n}})^2 \\
&\qquad\qquad \prod_{n < i < j \leq 2n} (-\sqrt{x_{i-n}} + \sqrt{x_{j-n}})^2 \\
&= \prod_{1 \leq i < j \leq n} (\sqrt{x_i} - \sqrt{x_j})^2 \prod_{\substack{1 \leq i \leq n \\ 1 \leq j \leq n}} (\sqrt{x_i} + \sqrt{x_j})^2
\end{aligned}
$$

$$\prod_{1 \le i < j \le n} (-\sqrt{x_i} + \sqrt{x_j})^2$$

$$= \prod_{1 \le i < j \le n} (\sqrt{x_i} - \sqrt{x_j})^4 \prod_{1 \le i < j \le n} (\sqrt{x_i} + \sqrt{x_j})^4$$

$$\prod_{1 \le i \le n} (2\sqrt{x_i})^2$$

$$= \prod_{1 \le i < j \le n} (x_i - x_j)^4 \ 2^{2n} \prod_{i=1}^{n} x_i$$

$$= 2^{2n} (-1)^n f(0) (D(f))^2 .$$

30.   Prove that for each positive integer n there exists a circle in the xy-plane which contains exactly n lattice points.

Solution: Let P be the point $(\sqrt{2}, 1/3)$. First, we show that two different lattice points $R = (x_1, y_1)$ and $S = (x_2, y_2)$ must be at different distances from P. For if R and S were at equal distances from P, then we would have

$$(x_1 - \sqrt{2})^2 + (y_1 - \frac{1}{3})^2 = (x_2 - \sqrt{2})^2 + (y_2 - \frac{1}{3})^2 ,$$

so that

$$(30.1) \qquad 2(x_2 - x_1)\sqrt{2} = x_2^2 + y_2^2 - x_1^2 - y_1^2 + \frac{2}{3}(y_1 - y_2) .$$

As $\sqrt{2}$ is irrational, from (30.1) we see that $x_1 - x_2 = 0$, and hence $y_2^2 - y_1^2 + \frac{2}{3}(y_1 - y_2) = 0$, that is

$$(y_2 - y_1)(y_2 + y_1 - 2/3) = 0 .$$

Since y_1 and y_2 are integers, we have $y_2 + y_1 - 2/3 \ne 0$, and so $y_1 = y_2$, contrary to the fact that R and S are assumed distinct.

Now let n be an arbitrary natural number. Let C be a circle with centre P and radius large enough so that C contains more than n lattice points. Clearly C contains a finite number m $(> n)$ of lattice points. As the distances from P to the lattice points are all different, we may arrange the lattice points inside C in a sequence $P_1, P_2, \ldots, P_m$, according to their increasing distances from P. Clearly, the circle C_n with centre P, passing through P_{n+1}, contains exactly n lattice points.

31. Let n be a given non-negative integer. Determine the number $S(n)$ of solutions of the equation

(31.0) $$x + 2y + 2z = n$$

in non-negative integers x, y, z.

Solution: We have for $|t| < 1$

$$
\begin{aligned}
\sum_{n=0}^{\infty} S(n)t^n &= (1 + t + t^2 + \cdots)(1 + t^2 + t^4 + \cdots)^2 \\
&= \frac{1}{(1-t)(1-t^2)^2} \\
&= \frac{1}{(1-t)^3(1+t)^2} \\
&= \frac{3/16}{1-t} + \frac{1/4}{(1-t)^2} + \frac{1/4}{(1-t)^3} + \frac{3/16}{1+t} + \frac{1/8}{(1+t)^2} \\
&= \frac{3}{16}\sum_{n=0}^{\infty} t^n + \frac{1}{4}\sum_{n=0}^{\infty}(n+1)t^n \\
&\quad + \frac{1}{4}\sum_{n=0}^{\infty}\frac{(n+1)(n+2)}{2}t^n + \frac{3}{16}\sum_{n=0}^{\infty}(-1)^n t^n \\
&\quad + \frac{1}{8}\sum_{n=0}^{\infty}(-1)^n(n+1)t^n \\
&= \frac{1}{16}\sum_{n=0}^{\infty}(3 + 4(n+1) + 2(n+1)(n+2) + 3(-1)^n
\end{aligned}
$$

$$+2(-1)^n(n+1))t^n \; ,$$

giving

$$S(n) = \begin{cases} \dfrac{n(n+6)}{8} + 1 & \text{, if } n \text{ is even ,} \\[3mm] \dfrac{(n+1)(n+3)}{8} & \text{, if } n \text{ is odd .} \end{cases}$$

32.   Let n be a fixed integer ≥ 2. Determine all functions $f(x)$, which are bounded for $0 < x < a$, and which satisfy the functional equation

$$(32.0) \qquad f(x) = \frac{1}{n^2} \left(f\left(\frac{x}{n}\right) + f\left(\frac{x+a}{n}\right) + \ldots + f\left(\frac{x+(n-1)a}{n}\right) \right).$$

Solution: Let $f(x)$ be a bounded function which satisfies (32.0) for $0 < x < a$. As $f(x)$ is bounded on $(0, a)$ there exists a positive constant K such that

$$(32.1) \qquad\qquad |f(x)| < K, \quad 0 < x < a \; .$$

For $s = 0, 1, \ldots, n-1$ we have

$$0 < \frac{x+sa}{n} < a, \quad \text{if } 0 < x < a \; ,$$

so that by (32.1) we obtain

$$\left| f\left(\frac{x+sa}{n}\right) \right| < K, \quad 0 \leq s \leq n-1, \quad 0 < x < a \; .$$

Then, for $0 < x < a$, we have from (32.0),

$$|f(x)| < \frac{1}{n^2}(K + K + \cdots + K),$$

that is $|f(x)| < K/n$. Repeating the argument with the bound K replaced by K/n, we obtain

$$|f(x)| < K/n^2, \quad 0 < x < a .$$

Continuing in this way we get

(32.2) $$|f(x)| < K/n^l, \quad 0 < x < a ,$$

for $l = 0, 1, \ldots$, and letting $l \to \infty$ in (32.2) gives $f(x) \equiv 0$ for $0 < x < a$.

33. Let I denote the closed interval $[a, b]$, $a < b$. Two functions $f(x)$, $g(x)$ are said to be *completely different* on I if $f(x) \neq g(x)$ for all x in I. Let $q(x)$ and $r(x)$ be functions defined on I such that the differential equation

$$\frac{dy}{dx} = y^2 + q(x)y + r(x)$$

has three solutions $y_1(x)$, $y_2(x)$, $y_3(x)$ which are pairwise completely different on I. If $z(x)$ is a fourth solution such that the pairs of functions $z(x)$, $y_i(x)$ are completely different for $i = 1, 2, 3$, prove that there exists a constant $K \ (\neq 0, 1)$ such that

(33.0) $$z = \frac{y_1(Ky_2 - y_3) + (1 - K)y_2y_3}{(K - 1)y_1 + (y_2 - Ky_3)} .$$

Solution: As $y_1, y_2, y_3, z = y_4$ are pairwise completely different on I, the function

(33.1) $$f(x) = \frac{(y_1 - y_2)(y_3 - y_4)}{(y_1 - y_3)(y_2 - y_4)}$$

is well-defined on I. Also, as y_1, y_2, y_3, y_4 are differentiable functions on I, $f(x)$ is differentiable there and its derivative is given by

$$f'(x) = \frac{g(x)}{(y_1 - y_3)^2(y_2 - y_4)^2} ,$$

where

$$g(x) = (y_1' - y_2')(y_1 - y_3)(y_2 - y_4)(y_3 - y_4)$$
$$- (y_1 - y_2)(y_1' - y_3')(y_2 - y_4)(y_3 - y_4)$$
$$- (y_1 - y_2)(y_1 - y_3)(y_2' - y_4')(y_3 - y_4)$$
$$+ (y_1 - y_2)(y_1 - y_3)(y_2 - y_4)(y_3' - y_4') \,.$$

As

$$y_s' = y_s^2 + qy_s + r, \quad s = 1, 2, 3, 4 \,,$$

we have

$$g(x) = ((y_1 + y_2 + q) - (y_1 + y_3 + q) - (y_2 + y_4 + q) + (y_3 + y_4 + q))$$
$$(y_1 - y_2)(y_1 - y_3)(y_2 - y_4)(y_3 - y_4) \,,$$

that is $g(x) = 0$, and so $f'(x) = 0$, showing that $f(x) = K$ on I for some constant K. Finally, (33.0) is obtained by solving (33.1) for $z = y_4$. $K \neq 0, 1$ as $z \neq y_3, y_1$ respectively.

34.   Let a_n, $n = 2, 3, \ldots$, denote the number of ways the product $b_1 b_2 \ldots b_n$ can be bracketed so that only two of the b_i are multiplied together at any one time. For example, $a_2 = 1$ since $b_1 b_2$ can only be bracketed as $(b_1 b_2)$, whereas $a_3 = 2$ as $b_1 b_2 b_3$ can be bracketed in two ways, namely, $(b_1(b_2 b_3))$ and $((b_1 b_2)b_3)$. Obtain a formula for a_n.

Solution: We set $a_1 = 1$. The number of ways of bracketing $b_1 b_2 \cdots b_{n+1}$ is

$$\sum_{i=1}^{n} N(1, i) \, N(i+1, n+1) \,,$$

where $N(i, j)$ denotes the number of ways of bracketing $b_i b_{i+1} \cdots b_j$, if $i < j$, and $N(i, j) = 1$, if $i = j$. Then

(34.1) $a_{n+1} = a_1 a_n + a_2 a_{n-1} + \cdots + a_{n-1} a_2 + a_n a_1, \quad n = 1, 2, \ldots \,.$

Set

$$(34.2) \qquad A(x) = \sum_{n=1}^{\infty} a_n x^n \ .$$

From (34.1) and (34.2) we obtain

$$
\begin{aligned}
A(x)^2 &= \left(\sum_{i=1}^{\infty} a_i x^i \right) \left(\sum_{j=1}^{\infty} a_j x^j \right) \\
&= \sum_{i,j=1}^{\infty} a_i a_j x^{i+j} = \sum_{n=1}^{\infty} \sum_{\substack{i,j=1 \\ i+j=n+1}}^{\infty} a_i a_j x^{i+j} \\
&= \sum_{n=1}^{\infty} (a_1 a_n + a_2 a_{n-1} + \cdots + a_n a_1) x^{n+1} \\
&= \sum_{n=1}^{\infty} a_{n+1} x^{n+1} = A(x) - x \ ,
\end{aligned}
$$

that is

$$(34.3) \qquad A(x)^2 - A(x) + x = 0 \ .$$

Solving the quadratic equation (34.3) for $A(x)$, we obtain

$$A(x) = (1 \pm \sqrt{1 - 4x})/2 \ .$$

As $A(0) = 0$ we must have

$$(34.4) \qquad A(x) = (1 - \sqrt{1 - 4x})/2 \ .$$

By the binomial theorem we have

$$(34.5) \qquad \sqrt{1 - 4x} = \sum_{n=0}^{\infty} \binom{1/2}{n} (-4)^n x^n \ ,$$

so that, from (34.4) and (34.5), we obtain

$$(34.6) \qquad A(x) = -\frac{1}{2} \sum_{n=1}^{\infty} \binom{1/2}{n} (-4)^n x^n \ .$$

Equating coefficients of x^n $(n \geq 2)$ in (34.6), we obtain

$$
\begin{aligned}
a_n &= -\frac{1}{2}\binom{1/2}{n}(-1)^n 2^{2n} \\
&= (-1)^{n-1} 2^{2n-1}\binom{1/2}{n} \\
&= (-1)^{n-1} 2^{2n-1}(-1)^{n-1}\frac{1.3.5\ldots(2n-3)}{2^n\, n!} ,
\end{aligned}
$$

that is

$$
a_n = \frac{1.3.5\ldots(2n-3)}{n!} 2^{n-1}, \quad n \geq 2 .
$$

35.   Evaluate the limit

(35.0) $$L = \lim_{y \to 0} \frac{1}{y}\int_0^\pi \tan(y \sin x)\, dx .$$

Solution: We begin by showing that

(35.1) $$t \leq \tan t \leq t + t^3, \quad 0 \leq t \leq 1 .$$

We set

$$f(t) = (\tan t - t)/t^3, \quad 0 < t \leq 1 ,$$

and deduce that

$$f'(t) = g(t)/t^4, \quad 0 < t \leq 1 ,$$

where

$$
\begin{cases}
g(t) &= t\tan^2 t - 3\tan t + 3t , \\
g'(t) &= \dfrac{\sin t}{\cos^3 t}(2t - \sin 2t) .
\end{cases}
$$

Hence $g'(t) > 0$, $0 < t \leq 1$, which implies that $g(t) > g(0) = 0$, $0 < t \leq 1$. We deduce that f is an increasing function on $0 < t \leq 1$, so that

$$f(0+) \leq f(t) \leq f(1), \quad 0 < t \leq 1 ,$$

that is

$$\frac{1}{3} \le \frac{\tan t - t}{t^3} \le \tan(1) - 1, \quad 0 < t \le 1 .$$

Since $\tan(1) < \tan(\pi/3) = \sqrt{3} < 2$, we have

$$0 < \frac{\tan t - t}{t^3} \le 1, \quad 0 < t \le 1 ,$$

which completes the proof of (35.1).

For $0 \le x \le \pi$ and $0 < y \le 1$ we have $0 \le \sin x \le 1$ and so

(35.2) $$0 \le y \sin x \le 1 .$$

Hence, by (35.1) and (35.2), we have

$$y \sin x \le \tan(y \sin x) \le y \sin x + (y \sin x)^3 ,$$

so that

(35.3) $$0 \le \frac{\tan(y \sin x) - y \sin x}{y} \le y^2 \sin^3 x .$$

Integrating (35.3) over $0 \le x \le \pi$, we obtain

(35.4) $$0 \le \frac{1}{y} \int_0^\pi (\tan(y \sin x) - y \sin x) \, dx \le y^2 \int_0^\pi \sin^3 x \, dx .$$

Letting $y \to 0+$ in (35.4) we deduce that

$$\lim_{y \to 0+} \frac{1}{y} \int_0^\pi (\tan(y \sin x) - y \sin x) \, dx = 0 ,$$

and thus

$$\lim_{y \to 0+} \frac{1}{y} \int_0^\pi \tan(y \sin x) \, dx = \int_0^\pi \sin x \, dx ,$$

that is

(35.5) $$\lim_{y \to 0+} \frac{1}{y} \int_0^\pi \tan(y \sin x) \, dx = 2 .$$

Replacing y by $-y$ in (35.5), we see that

(35.6) $$\lim_{y \to 0-} \frac{1}{y} \int_0^\pi \tan(y \sin x) \, dx = 2 \, ,$$

also. Hence, from (35.5) and (35.6), we find that $L = 2$.

36.   Let ϵ be a real number with $0 < \epsilon < 1$. Prove that there are infinitely many integers n for which

(36.0) $$\cos n \geq 1 - \epsilon \, .$$

Solution: According to a theorem of Hurwitz (1891): if θ is an irrational number, there are infinitely many rational numbers a/b with $b > 0$ and $GCD(a,b) = 1$ such that

$$\left| \theta - \frac{a}{b} \right| < \frac{1}{\sqrt{5} \, b^2} \, .$$

As π is irrational, Hurwitz's theorem implies that there are infinitely many rational numbers n/k with $k > 0$ and $GCD(n,k) = 1$ such that

$$\left| 2\pi - \frac{n}{k} \right| < \frac{1}{\sqrt{5} \, k^2} \, ,$$

or equivalently

(36.1) $$|2\pi k - n| < 1/(\sqrt{5} \, k) \, .$$

Let $0 < \epsilon < 1$. We consider those integers n and k satisfying (36.1) for which $k > 1/(\sqrt{5} \, \epsilon)$. There are clearly an infinite number of such positive integers k, and for each such k there is an integer n such that $|2\pi k - n| < \epsilon$. For such pairs (n, k) we have

$$1 - \cos n \quad \leq \quad |1 - \cos n|$$

$$= 2 \left| \sin \left(k\pi + \frac{n}{2} \right) \right| \left| \sin \left(k\pi - \frac{n}{2} \right) \right|$$

$$\leq 2 \left| \sin \left(k\pi - \frac{n}{2} \right) \right|$$

$$\leq 2 \left| k\pi - \frac{n}{2} \right|$$

$$= |2k\pi - n|$$

$$< \epsilon \,,$$

showing that (36.0) holds for infinitely many integers n.

37.   Determine all the functions f, which are everywhere differentiable and satisfy

$$(37.0) \qquad\qquad f(x) + f(y) = f \left(\frac{x + y}{1 - xy} \right)$$

for all real x and y with $xy \neq 1$.

Solution: Let $f(x)$ satisfy (37.0). Differentiating (37.0) partially with respect to each of x and y, we obtain

$$(37.1) \qquad\qquad f'(x) = \frac{1 + y^2}{(1 - xy)^2} \, f' \left(\frac{x + y}{1 - xy} \right)$$

and

$$(37.2) \qquad\qquad f'(y) = \frac{1 + x^2}{(1 - xy)^2} \, f' \left(\frac{x + y}{1 - xy} \right) \,.$$

Eliminating common terms in (37.1) and (37.2), we deduce that

$$(37.3) \qquad\qquad (1 + x^2) f'(x) = (1 + y^2) f'(y) \,.$$

As the left side of (37.3) depends only on x and the right side only on y, each side of (37.3) must be equal to a constant c. Thus we have

$$f'(x) = \frac{c}{1 + x^2} \,,$$

and so

$$f(x) = c \arctan x + d \ ,$$

for some constant d. However, taking $y = 0$ in (37.0), we obtain $f(x) + f(0) = f(x)$, so that $f(0) = 0$ and $d = 0$. Clearly $f(x) = c \arctan x$ satisfies (37.0), and so all solutions of (37.0) are given by

$$f(x) = c \arctan x \ ,$$

where c is a constant.

38.   A point X is chosen inside or on a circle. Two perpendicular chords AC and BD of the circle are drawn through X. (In the case when X is on the circle, the degenerate case, when one chord is a diameter and the other is reduced to a point, is allowed.) Find the greatest and least values which the sum $S = |AC| + |BD|$ can take for all possible choices of the point X.

Solution: We can choose an (x, y)-coordinate system in the plane so that the centre of the circle is at the origin, BD is parallel to the x-axis, AC is parallel to the y-axis, B lies to the left of D, and A lies above C. Let X denote a point (r, s) such that

$$(38.1) \qquad\qquad r^2 + s^2 \leq R^2 \ ,$$

where R is the radius of the circle. Then the coordinates of the points A, B, C, D are

$$(r, \sqrt{R^2 - r^2}), \quad (-\sqrt{R^2 - s^2}, s), \quad (r, -\sqrt{R^2 - r^2}), \quad (\sqrt{R^2 - s^2}, s)$$

respectively. Thus we have

$$|AC| = 2\sqrt{R^2 - r^2}, \quad |BD| = 2\sqrt{R^2 - s^2} \ ,$$

and so

$$S(r, s) = |AC| + |BD| = 2(\sqrt{R^2 - r^2} + \sqrt{R^2 - s^2}) \ .$$

We wish to find the maximum and minimum values of $S(r,s)$ subject to the constraint (38.1).

First we determine the maximum value of $S(r,s)$. Clearly, we have

$$\sqrt{R^2 - r^2} + \sqrt{R^2 - s^2} \leq 2R ,$$

and this proves that

$$\max_{r^2+s^2 \leq R^2} S(r,s) = S(0,0) = 4R .$$

Finally, we determine the minimum value of $S(r,s)$. We have

$$
\begin{aligned}
\left(\sqrt{R^2 - r^2} + \sqrt{R^2 - s^2}\right)^2 &= 2R^2 - (r^2 + s^2) + 2\sqrt{R^2 - r^2}\sqrt{R^2 - s^2} \\
&\geq 2R^2 - (r^2 + s^2) \\
&\geq 2R^2 - (r^2 + s^2) + (r^2 + s^2) - R^2 \\
&= R^2 ,
\end{aligned}
$$

so that

$$\sqrt{R^2 - r^2} + \sqrt{R^2 - s^2} \geq R .$$

This proves that

$$\min_{r^2+s^2 \leq R^2} S(r,s) = S(\pm R, 0) = S(0, \pm R) = 2R .$$

39.   For $n = 1, 2, \ldots$ define the set A_n by

$$
A_n = \begin{cases}
\{0, 2, 4, 6, 8, \ldots\}, & \text{if } n \equiv 0 \ (\mathrm{mod}\ 2) , \\
\{0, 3, 6, \ldots, 3(n-1)/2\}, & \text{if } n \equiv 1 \ (\mathrm{mod}\ 2) .
\end{cases}
$$

Is it true that

$$\bigcup_{n=1}^{\infty} \left(\bigcap_{k=1}^{\infty} A_{n+k} \right) = \bigcap_{n=1}^{\infty} \left(\bigcup_{k=1}^{\infty} A_{n+k} \right) ?$$

Solution: We set $X = \{0, 2, 4, 6, \dots \}$ and $Y = \{0, 3, 6, 9, \dots \}$. Clearly, we have

$$A_1 \subset A_3 \subset A_5 \subset \cdots \subset \bigcup_{n=0}^{\infty} A_{2n+1} = Y$$

and

$$A_2 = A_4 = A_6 = \cdots = X \ .$$

Hence, we have for $n = 1, 2, \dots$

$$\bigcap_{k=1}^{\infty} A_{n+k} = \bigcap_{\substack{k=1 \\ n+k \equiv 0 \ (\text{mod } 2)}}^{\infty} A_{n+k} \ \cap \ \bigcap_{\substack{k=1 \\ n+k \equiv 1 \ (\text{mod } 2)}}^{\infty} A_{n+k}$$
$$= X \cap B_n \ ,$$

where

$$B_n = \left\{ \begin{array}{ll} A_{n+1}, & \text{if } n \equiv 0 \pmod 2 \ , \\ A_{n+2}, & \text{if } n \equiv 1 \pmod 2 \ , \end{array} \right.$$

and so

$$\bigcup_{n=1}^{\infty} \left(\bigcap_{k=1}^{\infty} A_{n+k} \right) = \bigcup_{n=1}^{\infty} (X \cap B_n)$$
$$= X \cap \left(\bigcup_{n=1}^{\infty} B_n \right)$$
$$= X \cap \left(\bigcup_{n=1}^{\infty} A_{2n+1} \right)$$
$$= X \cap Y \ .$$

On the other hand, we have

$$\bigcup_{k=1}^{\infty} A_{n+k} = \bigcup_{\substack{k=1 \\ n+k \equiv 0 \ (\text{mod } 2)}}^{\infty} A_{n+k} \ \cup \ \bigcup_{\substack{k=1 \\ n+k \equiv 1 \ (\text{mod } 2)}}^{\infty} A_{n+k}$$
$$= X \cup Y$$

for all $n = 1, 2, \ldots$, so that

$$\bigcap_{n=1}^{\infty} \left(\bigcup_{k=1}^{\infty} A_{n+k} \right) = X \cup Y .$$

Hence, we see that

$$\bigcup_{n=1}^{\infty} \left(\bigcap_{k=1}^{\infty} A_{n+k} \right) \neq \bigcap_{n=1}^{\infty} \left(\bigcup_{k=1}^{\infty} A_{n+k} \right) ,$$

as 2 belongs to $X \cup Y$ but does not belong to $X \cap Y$.

40. A sequence of repeated independent trials is performed. Each trial has probability p of being successful and probability $q = 1 - p$ of failing. The trials are continued until an uninterrupted sequence of n successes is obtained. The variable X denotes the number of trials required to achieve this goal. If $p_k = \text{Prob}(X = k)$, determine the probability generating function $P(x)$ defined by

$$(40.0) \qquad\qquad P(x) = \sum_{k=0}^{\infty} p_k x^k .$$

Solution: Clearly, we have

$$p_k = \begin{cases} 0 & , k = 0, 1, \ldots, n - 1 , \\ p^n & , k = n , \\ qp^n & , k = (n + 1), (n + 2), \ldots, 2n . \end{cases}$$

For $k > 2n$ we have

$$p_k = \text{Prob}(A)\,\text{Prob}(B)\,\text{Prob}(C) ,$$

where A, B, C represent events as follows:

(A) no n consecutive successes in the first $k - n - 1$ trials;
(B) $(k - n)$ th trial is a failure;
(C) n successes in last n trials.

Then $p_k = (1 - \text{Prob}(D))qp^n$, where D represents the event of at least one run of n consecutive successes in the first $k - n - 1$ trials, that is

$$p_k = \left(1 - \sum_{i=0}^{k-n-1} p_i\right)qp^n, \quad k > 2n .$$

Hence we have

$$P(x) = p^n x^n + qp^n(x^{n+1} + \cdots + x^{2n}) + qp^n \sum_{k=2n+1}^{\infty}\left(1 - \sum_{i=0}^{k-n-1} p_i\right)x^k ,$$

and so

$$
\begin{aligned}
\frac{P(x)}{p^n x^n} &= 1 + q(x + \cdots + x^n) + q\sum_{k=2n+1}^{\infty} x^{k-n} - q\sum_{k=2n+1}^{\infty}\sum_{i=0}^{k-n-1} p_i x^{k-n} \\
&= 1 + q\frac{(x - x^{n+1})}{(1-x)} + q\frac{x^{n+1}}{(1-x)} - qx^{n+1}\sum_{l=0}^{\infty}\sum_{i=0}^{n+l} p_i x^l \\
&= \frac{(1 - x + qx)}{(1-x)} - qx^{n+1}\sum_{l=0}^{\infty}\sum_{i=n}^{n+l} p_i x^l \\
&= \frac{(1 - x + qx)}{(1-x)} - qx^{n+1}\sum_{l=0}^{\infty}\sum_{r=0}^{l} p_{n+r} x^l \\
&= \frac{(1 - x + qx)}{(1-x)} - qx^{n+1}\sum_{l=0}^{\infty}\sum_{\substack{r,s=0 \\ r+s=l}}^{l} p_{n+r} x^{r+s} \\
&= \frac{(1 - x + qx)}{(1-x)} - qx^{n+1}\sum_{r=0}^{\infty}\sum_{s=0}^{\infty} p_{n+r} x^{r+s} \\
&= \frac{(1 - x + qx)}{(1-x)} - qx\left(\sum_{r=0}^{\infty} p_{n+r} x^{n+r}\right)\left(\sum_{s=0}^{\infty} x^s\right) ,
\end{aligned}
$$

that is

$$\frac{P(x)}{p^n x^n} = \frac{(1 - x + qx)}{(1-x)} - qx\frac{P(x)}{(1-x)} ,$$

so that

$$P(x) = \frac{(1 - px)p^n x^n}{1 - x + qp^n x^{n+1}} .$$

41. A, B, C, D are four points lying on a circle such that $ABCD$ is a convex quadrilateral. Determine a formula for the radius of the circle in terms of $a = |AB|$, $b = |BC|$, $c = |CD|$ and $d = |DA|$.

Solution: We first prove the following result:
The radius of the circumcircle of a $\triangle LMN$ is given by

$$(41.1) \qquad R = \frac{lmn}{\sqrt{(l + m + n)(l + m - n)(l - m + n)(-l + m + n)}} \,,$$

where

$$l = |MN|, \quad m = |NL|, \quad n = |LM| \,.$$

Let C denote the circumcentre of $\triangle LMN$, so that $|LC| = |MC| = |NC| = R$.
Set

$$\alpha = \angle MCN, \quad \beta = \angle NCL, \quad \gamma = \angle LCM \,,$$

so that $\alpha + \beta + \gamma = 2\pi$. By the sine law applied to $\triangle MCN$ we have

$$\frac{l}{\sin \alpha} = \frac{R}{\sin((\pi - \alpha)/2)}$$

so that

$$l = R \frac{\sin \alpha}{\cos(\alpha/2)} = 2R \sin(\alpha/2) \,.$$

Similarly, we have

$$m = 2R \sin(\beta/2), \quad n = 2R \sin(\gamma/2) \,.$$

Thus we obtain

$$\begin{aligned} \frac{n}{2R} &= \sin(\gamma/2) \\ &= \sin\left(\pi - \frac{(\alpha + \beta)}{2}\right) \end{aligned}$$

$$= \sin\left(\frac{\alpha}{2} + \frac{\beta}{2}\right)$$

$$= \sin(\alpha/2)\cos(\beta/2) + \cos(\alpha/2)\sin(\beta/2)$$

$$= \frac{l}{2R}\sqrt{1 - \frac{m^2}{4R^2}} + \frac{m}{2R}\sqrt{1 - \frac{l^2}{4R^2}}$$

and so

$$n = l\sqrt{1 - \frac{m^2}{4R^2}} + m\sqrt{1 - \frac{l^2}{4R^2}} \, .$$

Squaring both sides we obtain

$$n^2 = l^2\left(1 - \frac{m^2}{4R^2}\right) + m^2\left(1 - \frac{l^2}{4R^2}\right) + 2lm\sqrt{\left(1 - \frac{l^2}{4R^2}\right)\left(1 - \frac{m^2}{4R^2}\right)} \, ,$$

and so

$$2lm\sqrt{\left(1 - \frac{l^2}{4R^2}\right)\left(1 - \frac{m^2}{4R^2}\right)} = (n^2 - l^2 - m^2) + \frac{l^2 m^2}{2R^2} \, .$$

Squaring again we find that

$$4l^2 m^2\left(1 - \frac{l^2}{4R^2}\right)\left(1 - \frac{m^2}{4R^2}\right) = (n^2 - l^2 - m^2)^2 + \frac{l^4 m^4}{4R^4}$$
$$+ \frac{l^2 m^2}{R^2}(n^2 - l^2 - m^2) \, ,$$

giving, after some simplification

$$(n^2 - l^2 - m^2)^2 - 4l^2 m^2 = -\frac{l^2 m^2 n^2}{R^2} \, ,$$

which establishes (41.1).

Returning to the original problem, we set $x = |AC|$, and $\theta = \angle ABC$, so that $\angle CDA = \pi - \theta$. By the cosine law in $\triangle ABC$ and $\triangle ACD$, we have

$$(41.2) \qquad\qquad x^2 = a^2 + b^2 - 2ab\cos\theta$$

and

(41.3) $x^2 = c^2 + d^2 - 2cd\cos(\pi - \theta) = c^2 + d^2 + 2cd\cos\theta$.

Eliminating x^2 from (41.2) and (41.3), we obtain

$$\cos\theta = \frac{a^2 + b^2 - c^2 - d^2}{2(ab + cd)} .$$

Using this expression for $\cos\theta$ in (41.2), we get

$$\begin{aligned} x^2 &= a^2 + b^2 - ab\frac{(a^2 + b^2 - c^2 - d^2)}{(ab + cd)} \\ &= \frac{(ac + bd)(ad + bc)}{(ab + cd)} , \end{aligned}$$

so that

$$x = \sqrt{\frac{(ac + bd)(ad + bc)}{(ab + cd)}} .$$

The radius r of the circle passing through A, B, C, D is the circumradius of $\triangle ABC$, and so by (41.1) is given by

$$\begin{aligned} r &= \frac{abx}{\sqrt{(a + b + x)(a + b - x)(a - b + x)(-a + b + x)}} \\ &= \frac{abx}{\sqrt{((a + b)^2 - x^2)(x^2 - (a - b)^2)}} . \end{aligned}$$

Next we have

$$\begin{aligned} (a + b)^2 - x^2 &= (a + b)^2 - \frac{(ac + bd)(ad + bc)}{(ab + cd)} \\ &= \frac{ab((a + b)^2 - (c - d)^2)}{(ab + cd)} \\ &= \frac{ab(a + b - c + d)(a + b + c - d)}{(ab + cd)} \end{aligned}$$

and similarly

$$x^2 - (a - b)^2 = \frac{ab(-a + b + c + d)(a - b + c + d)}{(ab + cd)} ,$$

so that

$$r = \sqrt{\frac{(ab + cd)(ac + bd)(ad + bc)}{(-a + b + c + d)(a - b + c + d)(a + b - c + d)(a + b + c - d)}} \ .$$

42.   Let $ABCD$ be a convex quadrilateral. Let P be the point outside $ABCD$ such that $|AP| = |PB|$ and $\angle APB = 90^0$. The points Q, R, S are similarly defined. Prove that the lines PR and QS are of equal length and perpendicular.

Solution: We consider the quadrilateral $ABCD$ to be in the complex plane and denote the vertices A, B, C, D by the complex numbers a, b, c, b. Then the midpoints H, K, L, M of the sides AB, BC, CD, DA are represented by $(a + b)/2, (b + c)/2, (c + d)/2, (d + a)/2$. Let p represent the point P. As $|PH| = |BH|$ and $PH \perp BH$ we have

$$p - \left(\frac{a + b}{2}\right) = i\left(b - \left(\frac{a + b}{2}\right)\right)$$

so that

$$p = \left(\frac{1 - i}{2}\right)(a + ib) \ .$$

Similarly, we find that

$$\begin{cases} q &= \left(\frac{1-i}{2}\right)(b + ic) \ , \\ r &= \left(\frac{1-i}{2}\right)(c + id) \ , \\ s &= \left(\frac{1-i}{2}\right)(d + ia) \ . \end{cases}$$

From this we obtain

$$\begin{aligned} p - r &= \left(\frac{1-i}{2}\right)((a - c) + i(b - d)) \ , \\ q - s &= \left(\frac{1-i}{2}\right)((b - d) + i(c - a)) \\ &= -i\left(\frac{1-i}{2}\right)((a - c) + i(b - d)) \ , \end{aligned}$$

so that $q - s = -i(p - r)$, proving that $|PR| = |QS|$ and $PR \perp QS$.

43. Determine polynomials $p(x, y, z, w)$ and $q(x, y, z, w)$ with real coefficients such that

$$(43.0) \qquad (xy + z + w)^2 - (x^2 - 2z)(y^2 - 2w)$$
$$\equiv (p(x, y, z, w))^2 - (x^2 - 2z)(q(x, y, z, w))^2 .$$

Solution: We seek a solution of (43.0) of the form

$$(43.1) \qquad \begin{cases} p(x, y, z, w) & = xy + X , \\ q(x, y, z, w) & = y + Y , \end{cases}$$

where X and Y are polynomials in x, w, and z. Substituting (43.1) in (43.0) and simplifying, we obtain

$$(43.2) \qquad \left((z - w)^2 + 2x^2 w \right) + 2x(z + w)y$$
$$\equiv \left(X^2 - (x^2 - 2z)Y^2 \right) + 2 \left(xX - (x^2 - 2z)Y \right) y ,$$

which gives

$$(43.3) \qquad \begin{cases} X^2 - (x^2 - 2z)Y^2 & = (z - w)^2 + 2x^2 w , \\ xX - (x^2 - 2z)Y & = x(z + w) . \end{cases}$$

From the second equation in (43.3) we have

$$X = \left((x^2 - 2z)Y + x(z + w) \right) / x ,$$

and, using this in the first equation in (43.3), we obtain after simplification

$$zY^2 - x(z + w)Y + x^2 w = 0 .$$

Solving for Y we find that $Y = xw/z$ or $Y = x$. Discarding the first solution as we are seeking polynomials X and Y, we have

$$X = x^2 - z + w, \quad Y = x ,$$

and so we may take

$$p(x, y, z, w) = xy + x^2 - z + w, \quad q(x, y, z, w) = x + y .$$

44.   Let $\mathbf{C}$ denote the field of complex numbers. Let $f : \mathbf{C} \to \mathbf{C}$ be a function satisfying

$$(44.0) \qquad \begin{cases} f(0) = 0 , \\ |f(z) - f(w)| = |z - w| , \end{cases}$$

for all z in $\mathbf{C}$ and $w = 0, 1, i$. Prove that

$$f(z) = f(1)z \quad \text{or} \quad f(1)\overline{z} ,$$

where $|f(1)| = 1$.

Solution: From (44.0) we have

$$(44.1) \qquad\qquad |f(z)| = |z| ,$$

$$(44.2) \qquad\qquad |f(z) - \alpha| = |z - 1| ,$$

$$(44.3) \qquad\qquad |f(z) - \beta| = |z - i| ,$$

which hold for all z in $\mathbf{C}$, and where

$$(44.4) \qquad\qquad \alpha = f(1), \quad \beta = f(i) .$$

Taking $z = 1, i$ in (44.1) and $z = i$ in (44.2), we obtain

$$(44.5) \qquad\qquad |\alpha| = |\beta| = 1, \quad |\alpha - \beta| = \sqrt{2} .$$

Hence, we have

$$
\begin{aligned}
\alpha^2 + \beta^2 &= \alpha^2 \beta \overline{\beta} + \alpha \overline{\alpha} \beta^2 \\
&= \alpha\beta(\alpha\overline{\beta} + \overline{\alpha}\beta) \\
&= \alpha\beta \left(\alpha\overline{\alpha} + \beta\overline{\beta} - (\alpha - \beta)(\overline{\alpha} - \overline{\beta}) \right) \\
&= \alpha\beta \left(|\alpha|^2 + |\beta|^2 - |\alpha - \beta|^2 \right) \\
&= \alpha\beta(1 + 1 - 2) \\
&= 0 ,
\end{aligned}
$$

so that

$$(44.6) \qquad\qquad \beta = \epsilon\alpha, \quad \epsilon = \pm i .$$

Next, squaring (44.2) and appealing to (44.1) and (44.5), we obtain

$$(44.7) \qquad\qquad \overline{\alpha} f(z) + \alpha \overline{f(z)} = z + \overline{z} ,$$

for all z in **C**. Similarly, squaring (44.3) and and appealing to (44.1), (44.5) and (44.6), we obtain

$$(44.8) \qquad\qquad \overline{\alpha} f(z) - \alpha \overline{f(z)} = -\epsilon i z + \epsilon i \overline{z} .$$

Adding (44.7) and (44.8), we deduce that

$$2\overline{\alpha} f(z) = (1 - \epsilon i)z + (1 + \epsilon i)\overline{z} ,$$

that is, as $\epsilon = \pm i$, $\overline{\alpha} f(z) = z$ or $\overline{z}$. Hence we have

$$f(z) = f(1)z \quad \text{or} \quad f(1)\overline{z} ,$$

where $|f(1)| = 1$, and it easy to check that both of these satisfy (44.0).

45. If x and y are rational numbers such that

$$(45.0) \qquad\qquad \tan \pi x = y ,$$

prove that $x = k/4$ for some integer k not congruent to 2 (mod 4).

Solution: As x and y are rational numbers there are integers p, q, r, s such that

$$\begin{cases} x = p/q, \quad y = r/s, \quad q > 0, \quad s > 0, \\ GCD(p,q) = GCD(r,s) = 1. \end{cases}$$

The equation (45.0) becomes

(45.1) $$\tan \pi \frac{p}{q} = \frac{r}{s}.$$

We have, appealing to DeMoivre's theorem,

$$\begin{aligned}
\left(\frac{s+ir}{s-ir} \right)^q &= \left(\frac{1+ir/s}{1-ir/s} \right)^q \\
&= \left(\frac{1+i\tan(\pi p/q)}{1-i\tan(\pi p/q)} \right)^q \\
&= \left(\frac{\cos(\pi p/q) + i\sin(\pi p/q)}{\cos(\pi p/q) - i\sin(\pi p/q)} \right)^q \\
&= \frac{\cos(\pi p) + i\sin(\pi p)}{\cos(\pi p) - i\sin(\pi p)} \\
&= \frac{(-1)^p + i.0}{(-1)^p - i.0} \\
&= 1,
\end{aligned}$$

so that, appealing to the binomial theorem, we have

$$\begin{aligned}
(s+ir)^q &= (s-ir)^q \\
&= ((s+ir) - 2ir)^q \\
&= \sum_{k=0}^{q} \binom{q}{k} (s+ir)^{q-k} (-2ir)^k.
\end{aligned}$$

Hence, we have

$$\begin{aligned}
(-2ir)^q &= -\sum_{k=1}^{q-1} \binom{q}{k} (s+ir)^{q-k} \\
&= -(s+ir) \sum_{k=1}^{q-1} \binom{q}{k} (s+ir)^{q-k-1},
\end{aligned}$$

that is

(45.2) $$(-2ir)^q = (s + ir)(x + iy),$$

for some integers x and y. Taking the modulus of both sides of (45.2), we obtain

$$2^{2q} r^{2q} = (s^2 + r^2)(x^2 + y^2).$$

Let p be an odd prime dividing $s^2 + r^2$. Then p divides $2^{2q} r^{2q}$ and so p divides r. Thus p divides $s^2 = (s^2 + r^2) - r^2$, that is, p divides s. This contradicts $GCD(r, s) = 1$. Thus $s^2 + r^2$ has no odd prime divisors and so must be a power of 2, say

$$s^2 + r^2 = 2^l, \quad l \geq 0.$$

Further, if $l \geq 2$, then s and r are both even, which is impossible, and so $l = 0$ or 1. As $s > 0$ we must have

$$(r, s) = (0, 1) \quad \text{or} \quad (\pm 1, 1).$$

The first possibility gives $x = k/4$, where $k = 4p$, while the second possibility gives $x = k/4$, where $k \equiv 1 \pmod 2$, thus completing the proof.

Second solution: (due to R. Dreyer) We make use of the fact that there are integers $c(n, r)$, $n = 1, 2, \ldots$; $r = 0, 1, \ldots, [n/2]$, such that

(45.3) $$2 \cos n\theta = \sum_{r=0}^{[n/2]} c(n, r)(2 \cos \theta)^{n-2r}$$

for any real number θ. The integers $c(n, r)$ are given recursively by

$$c(1, 0) = 1, \quad c(2, 0) = 1, \quad c(2, 1) = -2,$$

and for $n \geq 3$

$$\begin{cases} c(n, 0) &= 1, \\ c(n, r) &= c(n-1, r) - c(n-2, r-1), \quad 1 \leq r \leq (n-1)/2, \\ c(n, n/2) &= (-1)^{n/2} 2, \quad n \text{ even}. \end{cases}$$

Now, as x is rational, we may write $x = p/q$, where $GCD(p,q) = 1$ and $q > 0$. Further, as $y = \tan \pi x$ is rational, so is the quantity

$$z = 2 \cos 2\pi x = 2 \, \frac{(1 - \tan^2 \pi x)}{(1 + \tan^2 \pi x)} = 2 \, \frac{(1 - y^2)}{(1 + y^2)} \, .$$

Appealing to (45.3), with $n = q$ and $\theta = 2\pi x = 2\pi p/q$, we see that z is a rational root of the monic integral polynomial

$$f(x) = \sum_{r=0}^{[n/2]} c(n,r) x^{n-2r} - 2 \, .$$

Hence, z must be an integer. But $|z| = 2|\cos 2\pi x| \le 2$ so that $z = 0, \pm 1,$ or ± 2, that is

$$\cos(2\pi p/q) = 0, \pm 1/2, \pm 1 \, ,$$

giving

$$\frac{2\pi p}{q} = (2l + 1)\frac{\pi}{2}, \quad (3l \pm 1)\frac{\pi}{3}, \quad l\pi \, ,$$

for some integer l. Thus, we have

$$x = \frac{p}{q} = \frac{2l + 1}{4}, \quad \frac{3l \pm 1}{6}, \quad \text{or} \quad \frac{l}{2} \, .$$

Only the first possibility, and the third possibility with l even, have $y = \tan \pi x$ rational, and hence $x = k/4$, where k is not congruent to 2 (mod 4).

46.   Let P be a point inside the triangle ABC. Let AP meet BC at D, BP meet CA at E, and CP meet AB at F. prove that

(46.0) $$\frac{|PA|}{|PD|}\frac{|PB|}{|PE|} + \frac{|PB|}{|PE|}\frac{|PC|}{|PF|} + \frac{|PC|}{|PF|}\frac{|PA|}{|PD|} \ge 12 \, .$$

Solution: Let S, S_1, S_2, S_3 denote the areas of $\triangle ABC$, $\triangle PBC$, $\triangle PCA$, $\triangle PAB$ respectively, so that $S = S_1 + S_2 + S_3$. Since $\triangle ABC$ and

$\triangle PBC$ share the side BC, we have

$$\frac{|AD|}{|PD|} = \frac{S}{S_1} ,$$

so that

$$\begin{aligned}
\frac{|PA|}{|PD|} &= \frac{|AD| - |PD|}{|PD|} = \frac{|AD|}{|PD|} - 1 \\
&= \frac{S}{S_1} - 1 = \frac{S - S_1}{S_1} = \frac{S_2 + S_3}{S_1} .
\end{aligned}$$

Similarly, we have

$$\frac{|PB|}{|PE|} = \frac{S_3 + S_1}{S_2} , \quad \frac{|PC|}{|PF|} = \frac{S_1 + S_2}{S_3} .$$

Hence, we have

$$\begin{aligned}
&\frac{|PA|}{|PD|}\frac{|PB|}{|PE|} + \frac{|PB|}{|PE|}\frac{|PC|}{|PF|} + \frac{|PC|}{|PF|}\frac{|PA|}{|PD|} \\
&= \frac{(S_2 + S_3)(S_3 + S_1)}{S_1 S_2} + \frac{(S_3 + S_1)(S_1 + S_2)}{S_2 S_3} + \frac{(S_1 + S_2)(S_2 + S_3)}{S_3 S_1} \\
&= \left(\frac{S_3}{S_1} + \frac{S_3}{S_2} + 1 + \frac{S_3^2}{S_1 S_2} \right) + \left(\frac{S_1}{S_2} + \frac{S_1}{S_3} + 1 + \frac{S_1^2}{S_2 S_3} \right) \\
&\quad + \left(\frac{S_2}{S_3} + \frac{S_2}{S_1} + 1 + \frac{S_2^2}{S_3 S_1} \right) \\
&= \left(\frac{S_3}{S_1} + \frac{S_1}{S_3} \right) + \left(\frac{S_3}{S_2} + \frac{S_2}{S_3} \right) + \left(\frac{S_1}{S_2} + \frac{S_2}{S_1} \right) \\
&\quad + 3 + \left(\frac{S_1^2}{S_2 S_3} + \frac{S_2^2}{S_3 S_1} + \frac{S_3^2}{S_1 S_2} \right) \\
&\geq 2 + 2 + 2 + 3 + 3 = 12 ,
\end{aligned}$$

by the arithmetic-geometric mean inequality, which completes the proof of (46.0).

47.   Let l and n be positive integers such that

$$1 \leq l < n, \quad GCD(l, n) = 1 .$$

Define the integer k uniquely by

$$1 \leq k < n, \quad kl \equiv -1 \pmod{n} .$$

Let M be the $k \times l$ matrix whose (i,j)-th entry is

$$(i-1)l + j .$$

Let N be the $k \times l$ matrix formed by taking the columns of M in reverse order and writing the entries as the rows of N. What is the relationship between the (i,j)-th entry of M and the (i,j)-th entry of N modulo n?

Solution: If $A = [a_{ij}]$ and $B = [b_{ij}]$ are two $k \times l$ matrices, we write $A \equiv B \pmod{n}$ if $a_{ij} \equiv b_{ij} \pmod{n}$, $i = 1, 2, \ldots, k$; $j = 1, 2, \ldots, l$. As $kl \equiv -1 \pmod{n}$ we have modulo n

$$M = \begin{bmatrix} 1 & 2 & \cdots & l-2 & l-1 & l \\ l+1 & l+2 & \cdots & 2l-2 & 2l-1 & 2l \\ 2l+1 & 2l+2 & \cdots & 3l-2 & 3l-1 & 3l \\ \vdots & \vdots & \ddots & \vdots & \vdots & \vdots \\ (k-1)l+1 & (k-1)l+2 & \cdots & kl-2 & kl-1 & kl \end{bmatrix}$$

$$\equiv \begin{bmatrix} (kl-(k-1))l & (kl-(2k-1))l & \cdots & (2k+1)l & (k+1)l & l \\ (kl-(k-2))l & (kl-(2k-2))l & \cdots & (2k+2)l & (k+2)l & 2l \\ (kl-(k-3))l & (kl-(2k-3))l & \cdots & (2k+3)l & (k+3)l & 3l \\ \vdots & \vdots & \ddots & \vdots & \vdots & \vdots \\ (kl)l & (kl-k)l & \cdots & 3kl & 2kl & kl \end{bmatrix}$$

from which it is clear that the (i,j)-th entry of N is l times the (i,j)-th entry of M modulo n.

48. Let m and n be integers such that $1 \leq m < n$. Let a_{ij}, $i = 1, 2, \ldots, m$; $j = 1, 2, \ldots, n$, be mn integers which are not all zero, and set

$$a = \max_{\substack{1 \leq i \leq m \\ 1 \leq j \leq n}} |a_{ij}| .$$

Prove that the system of equations

$$(48.0) \quad \begin{cases} a_{11}x_1 + a_{12}x_2 + \cdots + a_{1n}x_n &= 0 \,, \\ a_{21}x_1 + a_{22}x_2 + \cdots + a_{2n}x_n &= 0 \,, \\ \vdots \\ a_{m1}x_1 + a_{m2}x_2 + \cdots + a_{mn}x_n &= 0 \,, \end{cases}$$

has a solution in integers $x_1, x_2, \ldots, x_n$, not all zero, satisfying

$$|x_j| \le \left[(2na)^{\frac{m}{n-m}} \right] \,, \quad 1 \le j \le n \,.$$

Solution: We set

$$N = \left[(2na)^{\frac{m}{n-m}} \right] \,,$$

so that

$$N > (2na)^{\frac{m}{n-m}} - 1, \quad \text{which implies} \quad (N+1)^{n-m} > (2na)^m \,.$$

Hence, we have

$$\begin{aligned} (N+1)^n &> (2na)^m (N+1)^m \\ &= (2naN + 2na)^m \,, \end{aligned}$$

that is, as $a \ge 1$,

$$(48.1) \qquad\qquad (N+1)^n > (2naN + 1)^m \,.$$

Set

$$L_i = L_i(y_1, y_2, \ldots, y_n) = a_{i1}y_1 + a_{i2}y_2 + \cdots + a_{in}y_n \,,$$

for $1 \le i \le m$. If $(y_1, y_2, \ldots, y_n)$ is a vector of integers satisfying $0 \le y_j \le N$, $1 \le j \le n$, the corresponding value of $L_i = L_i(y_1, y_2, \ldots, y_n)$, $1 \le i \le m$, satisfies

$$-naN \le L_i \le naN \,, \quad 1 \le i \le m \,,$$

and so the vector $(L_1, L_2, \ldots, L_m)$ of integers can take on at most $(2naN + 1)^m$ different values. As there are $(N + 1)^n$ choices of the vector $(y_1, y_2, \ldots, y_n)$, by (48.1) there must be two distinct vectors

$$\underline{u} = (y_1, y_2, \ldots, y_n), \qquad \underline{v} = (z_1, z_2, \ldots, z_n),$$

say, giving rise to the same vector $(L_1, L_2, \ldots, L_m)$. Set

$$x_j = y_j - z_j, \quad 1 \le j \le n.$$

As the two vectors $\underline{u}$ and $\underline{v}$ are distinct, not all the x_j are zero. Moreover, as

$$L_i(y_1, y_2, \ldots, y_n) = L_i(z_1, z_2, \ldots, z_n), \quad 1 \le i \le m,$$

$(x_1, x_2, \ldots, x_n)$ is a solution of (48.0). Finally, $|x_j| \le N$, $1 \le j \le n$, follows from the fact that $0 \le y_j, z_j \le N$, $1 \le j \le n$.

49.   Liouville proved that if

$$\int f(x) e^{g(x)} dx$$

is an elementary function, where $f(x)$ and $g(x)$ are rational functions with degree of $g(x) > 0$, then

$$\int f(x) e^{g(x)} dx = h(x) e^{g(x)},$$

where $h(x)$ is a rational function. Use Liouville's result to prove that

$$\int e^{-x^2} dx$$

is not an elementary function.

Solution: Suppose that $\int e^{-x^2} dx$ is an elementary function. Then, by Liouville's result, there exists a rational function $h(x)$ such that

$$\int e^{-x^2} dx = h(x) e^{-x^2}.$$

Hence, we have

$$\frac{d}{dx}(h(x)e^{-x^2}) = e^{-x^2} ,$$

and so

(49.1) $h'(x) - 2xh(x) = 1 .$

As $h(x)$ is a rational function we may write

(49.2) $h(x) = \frac{p(x)}{q(x)} ,$

where $p(x)$ and $q(x)$ are polynomials with $q(x)$ not identically zero, and $GCD(p(x), q(x)) = 1$. Then

(49.3) $h'(x) = \frac{p'(x)q(x) - p(x)q'(x)}{q(x)^2} ,$

and using (49.2) and (49.3) in (49.1), we obtain

(49.4) $p'(x)q(x) - p(x)q'(x) - 2xp(x)q(x) = q(x)^2 .$

If $q(x)$ is a constant polynomial, say $q(x) \equiv k$, then (49.4) becomes

$$p'(x) - 2xp(x) = k ,$$

which is clearly impossible as the degree of the polynomial on the left side is at least one. Thus, $q(x)$ is a non-constant polynomial. Let c denote one of its (complex) roots, and let m (≥ 1) denote the multiplicity of c so that $(x - c)^m \parallel q(x)$. Then, we have $(x - c)^{m-1} \parallel q'(x)$, and from (49.4) written in the form

$$p(x)q'(x) = (p'(x) - 2xp(x) - q(x))q(x) ,$$

we see that $(x - c) \mid p(x)$, which contradicts $GCD(p(x), q(x)) = 1$, and completes the proof.

50.   The sequence $x_0, x_1, \ldots$ is defined by the conditions

(50.0) $x_0 = 0, \quad x_1 = 1, \quad x_{n+1} = \frac{x_n + nx_{n-1}}{n+1}, \quad n \geq 1 .$

Determine

$$L = \lim_{n \to \infty} x_n .$$

Solution: The recurrence relation can be written as

$$x_{n+1} - x_n = -\frac{n}{n+1}(x_n - x_{n-1}), \quad n \geq 1 ,$$

so that

(50.1) $\qquad x_{n+1} - x_n = (-1)^n \frac{1}{n+1}(x_1 - x_0) = \frac{(-1)^n}{n+1}, \quad n \geq 1 .$

The equation in (50.1) trivially holds for $n = 0$. Hence, for $N \geq 1$, we have

$$
\begin{aligned}
x_N &= \sum_{n=0}^{N-1} (x_{n+1} - x_n) \\
&= \sum_{n=0}^{N-1} \frac{(-1)^n}{n+1} ,
\end{aligned}
$$

and so

$$
\begin{aligned}
L = \lim_{N \to \infty} x_N &= \lim_{N \to \infty} \sum_{n=0}^{N-1} \frac{(-1)^n}{n+1} \\
&= \sum_{n=0}^{\infty} \frac{(-1)^n}{n+1} ,
\end{aligned}
$$

that is $L = \ln 2$.

51.   Prove that the only integers $N \geq 3$ with the following property:

(51.0)   if $1 < k \leq N$ and $GCD(k, N) = 1$ then k is prime,

are

$$N = 3, 4, 6, 8, 12, 18, 24, 30 .$$

Solution: It is easy to check that $3, 4, 6, 8, 12, 18, 24, 30$ are the only integers ≤ 121 with the given property. Suppose that $N > 121$ is an integer with the property (51.0). Define the positive integer $n \geq 5$ by

$$(51.1) \qquad p_n \leq \sqrt{N} < p_{n+1} ,$$

where p_k denotes the k-th prime. From (51.1) we see that $p_j^2 \leq N$, $j = 1, 2, \ldots, n$, and so by property (51.0) we must have $p_j \mid N$, for $j = 1, 2, \ldots, n$. As $p_1, \ldots, p_n$ are distinct primes, we must have

$$(51.2) \qquad p_1 p_2 \cdots p_n \mid N ,$$

and so, by (51.1) and (51.2), we have

$$(51.3) \qquad p_1 p_2 \cdots p_n \leq N < p_{n+1}^2 .$$

By Bertrand's postulate, we have

$$p_{n+1} \leq 2p_n, \quad p_n \leq 2p_{n-1} ,$$

and so

$$(51.4) \qquad p_{n-1} p_n \geq \frac{p_n^2}{2} \geq \frac{p_{n+1}^2}{8} .$$

Using the inequality (51.4) in (51.3), we obtain

$$p_1 p_2 \cdots p_{n-2} p_{n+1}^2 / 8 < p_{n+1}^2 ,$$

that is $p_1 p_2 \cdots p_{n-2} < 8$. Since $p_1 p_2 = 6$, and $p_1 p_2 p_3 = 30$, we must have $n - 2 \leq 2$, and $n \leq 4$, which is impossible, proving that there are no integers $N > 121$ with property (51.0).

52. Find the sum of the infinite series

$$S = 1 - \frac{1}{4} + \frac{1}{6} - \frac{1}{9} + \frac{1}{11} - \frac{1}{14} + \cdots .$$

Solution: We begin by observing that

$$S = \frac{1}{1} - \frac{1}{4} + \frac{1}{6} - \frac{1}{9} + \frac{1}{11} - \frac{1}{14} + \cdots$$

$$= \int_0^1 (1 - x^3 + x^5 - x^8 + x^{10} - \cdots)\, dx$$

$$= \int_0^1 (1 - x^3)(1 + x^5 + x^{10} + \cdots)\, dx$$

$$= \int_0^1 \frac{1 - x^3}{1 - x^5}\, dx$$

$$= \int_0^1 \frac{x^2 + x + 1}{x^4 + x^3 + x^2 + x + 1}\, dx \ .$$

Now, decomposing into partial fractions, we have

$$\frac{x^2 + x + 1}{x^4 + x^3 + x^2 + x + 1} \equiv \frac{a}{x^2 + cx + 1} + \frac{b}{x^2 + dx + 1} \ ,$$

where

$$a = \frac{5 + \sqrt{5}}{10}, \quad b = \frac{5 - \sqrt{5}}{10},$$
$$c = \frac{1 - \sqrt{5}}{2}, \quad d = \frac{1 + \sqrt{5}}{2} \ .$$

Thus, we have

$$S = aI_c + bI_d \ ,$$

where

$$I_c = \int_0^1 \frac{dx}{x^2 + cx + 1}, \qquad I_d = \int_0^1 \frac{dx}{x^2 + dx + 1} \ .$$

Now

$$\int \frac{dx}{x^2 + 2tx + 1} = \frac{1}{\sqrt{1 - t^2}} \arctan\left(\frac{x + t}{\sqrt{1 - t^2}}\right), \quad |t| < 1 \ ,$$

and by the fundamental theorem of calculus, we have

$$\int_0^1 \frac{dx}{x^2 + 2tx + 1} = \frac{1}{\sqrt{1 - t^2}}\left(\arctan\left(\frac{1 + t}{\sqrt{1 - t^2}}\right) - \arctan\left(\frac{t}{\sqrt{1 - t^2}}\right)\right)$$

$$= \frac{1}{\sqrt{1 - t^2}} \arctan\left(\sqrt{\frac{1 - t}{1 + t}}\right) \ .$$

Hence, taking $t = (1 - \sqrt{5})/4$ and $t = (1 + \sqrt{5})/4$, we obtain

$$I_c = \sqrt{\frac{10 - 2\sqrt{5}}{5}} \arctan\left(\frac{\sqrt{5 + 2\sqrt{5}}}{\sqrt{5}}\right)$$

and

$$I_d = \sqrt{\frac{10 + 2\sqrt{5}}{5}} \arctan\left(\frac{\sqrt{5 - 2\sqrt{5}}}{\sqrt{5}}\right) .$$

Now

$$\cos(\pi/10) = (\sqrt{10 + 2\sqrt{5}})/4 , \quad \sin(\pi/10) = (\sqrt{5} - 1)/4 ,$$
$$\cos(3\pi/10) = (\sqrt{10 - 2\sqrt{5}})/4 , \quad \sin(3\pi/10) = (\sqrt{5} + 1)/4 ,$$

so that

$$\tan(\pi/10) = \frac{\sqrt{5 - 2\sqrt{5}}}{\sqrt{5}}, \qquad \tan(3\pi/10) = \frac{\sqrt{5 + 2\sqrt{5}}}{\sqrt{5}} .$$

Hence, we find that

$$I_c = \frac{3\pi}{10}\sqrt{\frac{10 - 2\sqrt{5}}{5}}, \qquad I_d = \frac{\pi}{10}\sqrt{\frac{10 + 2\sqrt{5}}{5}} ,$$

and so

$$\begin{aligned}
S &= \frac{\pi}{100}\left(3(5 + \sqrt{5})\sqrt{\frac{10 - 2\sqrt{5}}{5}} + (5 - \sqrt{5})\sqrt{\frac{10 + 2\sqrt{5}}{5}}\right) \\
&= \frac{\pi}{100}\left(3(\sqrt{5} + 1)\sqrt{10 - 2\sqrt{5}} + (\sqrt{5} - 1)\sqrt{10 + 2\sqrt{5}}\right) \\
&= \frac{\pi}{100}\left(6\sqrt{10 + 2\sqrt{5}} + 2\sqrt{10 - 2\sqrt{5}}\right) \\
&= \frac{\pi}{50}\left(3\sqrt{10 + 2\sqrt{5}} + \sqrt{10 - 2\sqrt{5}}\right) ,
\end{aligned}$$

as required.

53.   Semicircles are drawn externally to the sides of a given triangle. The lengths of the common tangents to these semicircles are l, m, and n. Relate the quantity

$$\frac{lm}{n} + \frac{mn}{l} + \frac{nl}{m}$$

to the lengths of the sides of the triangle.

Solution: Let the vertices of the given triangle be A, B, C. Let A', B', C' be the centres of the semicircles α, β, γ drawn on BC, CA, AB respectively. Let DE, FG, HJ be the common tangents to β and γ, γ and α, α and β respectively. Join $B'D, C'E$ and draw $C'K$ from C' perpendicular to $B'D$. Hence, as $KC'ED$ is a rectangle, we have $KC' = DE = l$. Let

$$|AB| = 2c, \quad |BC| = 2a, \quad |CA| = 2b \,.$$

Then, we have

$$|B'C'| = a, \quad |B'K| = |b - c| \,,$$

and so

$$|KC'| = \sqrt{a^2 - (b - c)^2} \,,$$

that is

$$l = \sqrt{(a - b + c)(a + b - c)} \,.$$

Similarly, we have

$$\begin{cases} m &= |FG| &= \sqrt{(a + b - c)(-a + b + c)} \,, \\ n &= |HJ| &= \sqrt{(-a + b + c)(a - b + c)} \,, \end{cases}$$

and so

$$\frac{mn}{l} = -a + b + c, \quad \frac{nl}{m} = a - b + c, \quad \frac{lm}{n} = a + b - c \,,$$

giving

(53.1) $$\frac{mn}{l} + \frac{nl}{m} + \frac{lm}{n} = a + b + c \,,$$

so that the left side of (53.1) is the semiperimeter of the triangle.

54. Determine all the functions $H : \mathbf{R}^4 \to \mathbf{R}$ having the properties

(i) $H(1,0,0,1) = 1$,
(ii) $H(\lambda a, b, \lambda c, d) = \lambda H(a,b,c,d)$,
(iii) $H(a,b,c,d) = -H(b,a,d,c)$,
(iv) $H(a+e, b, c+f, d) = H(a,b,c,d) + H(e,b,f,d)$,

where $a, b, c, d, e, f, \lambda$ are real numbers.

Solution: By (iii) we have

$$H(1,1,0,0) = -H(1,1,0,0), \quad H(0,0,1,1) = -H(0,0,1,1),$$

so that

$$(54.1) \qquad H(1,1,0,0) = H(0,0,1,1) = 0,$$

and from (i) and (iii) we have

$$(54.2) \qquad H(0,1,1,0) = -H(1,0,0,1) = -1.$$

Hence, we obtain

$$
\begin{aligned}
H(a,b,c,d) &= H(a,b,0,d) + H(0,b,c,d) &&\text{(by } (iv)) \\
&= aH(1,b,0,d) + cH(0,b,1,d) &&\text{(by } (ii)) \\
&= -aH(b,1,d,0) - cH(b,0,d,1) &&\text{(by } (iii)) \\
&= -a\left(H(b,1,0,0) + H(0,1,d,0)\right) \\
&\quad -c\left(H(b,0,0,1) + H(0,0,d,1)\right) &&\text{(by } (iv)) \\
&= -abH(1,1,0,0) - adH(0,1,1,0) \\
&\quad -bcH(1,0,0,1) - cdH(0,0,1,1) &&\text{(by } (ii)) \\
&= -ab(0) - ad(-1) - bc(1) - cd(0) \\
&= ad - bc,
\end{aligned}
$$

that is

$$H(a,b,c,d) = \begin{vmatrix} a & b \\ c & d \end{vmatrix}.$$

55. Let $z_1, \ldots, z_n$ be the complex roots of the equation

$$z^n + a_1 z^{n-1} + \ldots + a_n = 0 \,,$$

where $a_1, \ldots, a_n$ are n (≥ 1) complex numbers. Set

$$A = \max_{1 \leq k \leq n} |a_k| \,.$$

Prove that

$$|z_j| \leq 1 + A, \quad j = 1, 2, \ldots, n.$$

Solution: Set

$$f(z) = z^n + a_1 z^{n-1} + \cdots + a_n$$

and suppose that one of the z_j, $1 \leq j \leq n$, is such that $|z_j| > 1 + A$. Then, we have

$$
\begin{aligned}
0 = |f(z_j)| \;&=\; \left| z_j^n \left(1 + \frac{a_1}{z_j} + \cdots + \frac{a_n}{z_j^n} \right) \right| \\
&=\; |z_j|^n \left| 1 + \frac{a_1}{z_j} + \cdots + \frac{a_n}{z_j^n} \right| \\
&\geq\; |z_j|^n \left(1 - \frac{|a_1|}{|z_j|} - \cdots - \frac{|a_n|}{|z_j|^n} \right) \\
&\geq\; |z_j|^n \left(1 - \frac{A}{|z_j|} - \cdots - \frac{A}{|z_j|^n} \right) \\
&\geq\; |z_j|^n \left(1 - \frac{A}{|z_j|} - \cdots - \frac{A}{|z_j|^n} - \cdots \right) \\
&=\; |z_j|^n \left(1 - \frac{A}{|z_j| - 1} \right) \\
&=\; |z_j|^n \frac{(|z_j| - (A+1))}{|z_j| - 1} \\
&>\; 0 \,,
\end{aligned}
$$

which is impossible. Thus all the roots z_j, $1 \leq j \leq n$, of $f(z)$ must satisfy $|z_j| \leq 1 + A$.

56.   If m and n are positive integers with m odd, determine

$$d = GCD(2^m - 1, 2^n + 1) .$$

Solution: Define integers k and l by

$$2^m - 1 = kd, \quad 2^n + 1 = ld ,$$

and then we obtain

$$2^m = kd + 1, \quad 2^n = ld - 1 ,$$

and so for integers s and t we have

$$\begin{cases} 2^{mn} = (kd + 1)^n = sd + 1 \\ 2^{mn} = (ld - 1)^m = td - 1 , & \text{as } m \text{ is odd.} \end{cases}$$

Hence, we have $(s - t)d = -2$, and so d divides 2. But clearly d is odd, so that $d = 1$.

57.   If $f(x)$ is a polynomial of degree $2m + 1$ with integral coefficients for which there are $2m + 1$ integers $k_1, \ldots, k_{2m+1}$ such that

(57.0) $f(k_1) = \ldots = f(k_{2m+1}) = 1 ,$

prove that $f(x)$ is not the product of two non-constant polynomials with integral coefficients.

Solution: Suppose that $f(x)$ is the product of two non-constant polynomials with integral coefficients, say

$$f(x) = g(x)h(x) ,$$

where $r = \deg(g(x))$ and $s = \deg(h(x))$ satisfy

$$r + s = 2m + 1, \qquad 1 \le r \le s \le 2m \,.$$

Clearly, we have $r \le m$. Now, for $i = 1, 2, \ldots, 2m + 1$, we have, from (57.0),

$$1 = f(k_i) = g(k_i)h(k_i) \,.$$

As $g(k_i)$ is an integer, we must have

$$g(k_i) = \pm 1, \qquad i = 1, 2, \ldots, 2m + 1 \,.$$

Clearly, either $+1$ or -1 occurs at least $m + 1$ times among the values of $g(k_i)$, $1 \le i \le 2m + 1$, and we let ϵ denote this value. Then $g(x) - \epsilon$ is a polynomial of degree at most m which vanishes for at least $m + 1$ values of x. Hence the polynomial $g(x) - \epsilon$ must vanish identically, that is, $g(x)$ is a constant polynomial, which is a contradiction. Thus there is no factorization of $f(x)$ of the type supposed.

       58.   Prove that there do not exist integers a, b, c, d (not all zero) such that

(58.0) $$a^2 + 5b^2 - 2c^2 - 2cd - 3d^2 = 0 \,.$$

Solution: Suppose that (58.0) has a solution in integers a, b, c, d which are not all zero. Set

$$\begin{cases} m = GCD(a, b, c, d) \,, \\ a_1 = a/m, \quad b_1 = b/m, \quad c_1 = c/m, \quad d_1 = d/m \,. \end{cases}$$

Then clearly (a_1, b_1, c_1, d_1) is a solution in integers, not all zero, of (58.0) with

$$GCD(a_1, b_1, c_1, d_1) = 1 \,.$$

Hence we may suppose, without loss of generality, that (a, b, c, d) is a solution of (58.0) with $GCD(a, b, c, d) = 1$. Then, from (58.0), we obtain

(58.1) $$2(a^2 + 5b^2) = (2c + d)^2 + 5d^2 ,$$

so that $2a^2 \equiv (2c + d)^2 \pmod 5$. Since 2 is a quadratic nonresidue (mod 5) we must have

(58.2) $$a \equiv 2c + d \equiv 0 \pmod 5 .$$

Set
$$a = 5X, \qquad 2c + d = 5Y ,$$

where X and Y are integers, so that (58.1) becomes

$$2(5X^2 + b^2) = 5Y^2 + d^2 .$$

Thus we have $2b^2 \equiv d^2 \pmod 5$. Again, as 2 is a quadratic nonresidue (mod 5), we deduce that

(58.3) $$b \equiv d \equiv 0 \pmod 5 .$$

Appealing to (58.2) and (58.3), we see that $a \equiv b \equiv c \equiv d \equiv 0 \pmod 5$, contradicting $GCD(a, b, c, d) = 1$. Hence the only solution of (58.0) in integers is $a = b = c = d = 0$.

59. Prove that there exist infinitely many positive integers which are not representable as sums of fewer than ten squares of odd natural numbers.

Solution: We show that the positive integers $72k + 42$, $k = 0, 1, \ldots$, cannot be expressed as sums of fewer than ten squares of odd natural numbers. For suppose that

(59.1) $$72k + 42 = x_1^2 + x_2^2 + \cdots + x_s^2 ,$$

for some $k \geq 0$, where $x_1, \ldots, x_s$ are odd integers and $1 \leq s < 10$. Now, $x_i^2 \equiv 1 \pmod 8$ for $i = 1, 2, \ldots, s$, and so considering (59.1) as a congruence modulo 8, we have
$$s \equiv 2 \pmod 8 .$$

Since $1 \le s < 10$ we must have $s = 2$ and so

$$(59.2) \qquad\qquad 72k + 42 = x_1^2 + x_2^2 \ .$$

Treating (59.2) as a congruence modulo 3, we obtain

$$x_1^2 + x_2^2 \equiv 0 \pmod 3 \ .$$

Since the square of an integer is congruent to 0 or 1 (mod 3), we must have $x_1 \equiv x_2 \equiv 0 \pmod 3$. Finally, reducing (59.2) modulo 9, we obtain the contradiction $6 \equiv 0 \pmod 9$.

60.   Evaluate the integral

$$(60.0) \qquad\qquad I(k) = \int_0^\infty \frac{\sin kx \cos^k x}{x} \, dx \ ,$$

where k is a positive integer.

Solution: By the binomial theorem, we have

$$(60.1) \qquad\qquad (e^{2ix} + 1)^k = \sum_{r=0}^k \binom{k}{r} e^{2rix} \ .$$

As

$$(e^{2ix} + 1)^k = e^{kix}(e^{ix} + e^{-ix})^k = (\cos kx + i \sin kx)2^k \cos^k x \ ,$$

the imaginary part of $(e^{2ix} + 1)^k$ is $2^k \sin kx \cos^k x$. Equating imaginary parts in (60.1), we obtain

$$2^k \sin kx \cos^k x = \sum_{r=0}^k \binom{k}{r} \sin 2rx = \sum_{r=1}^k \binom{k}{r} \sin 2rx \ .$$

Thus, using $\int_0^\infty \frac{\sin x}{x} \, dx = \frac{\pi}{2}$, we have

$$I(k) \quad = \quad \frac{1}{2^k} \sum_{r=1}^k \binom{k}{r} \int_0^\infty \frac{\sin 2rx}{x} \, dx$$

$$= \frac{\pi}{2^{k+1}} \sum_{r=1}^{k} \binom{k}{r}$$

$$= \frac{\pi}{2^{k+1}} (2^k - 1)$$

$$= \frac{\pi}{2} \left(1 - \frac{1}{2^k} \right) ,$$

as required.

61. Prove that

$$\frac{1}{n+1} \binom{2n}{n}$$

is an integer for $n = 1, 2, 3, \dots$.

Solution: For $n = 1, 2, \dots$, we have

$$\frac{1}{n+1} \binom{2n}{n} = \frac{2n!}{(n!)^2} \frac{1}{n+1}$$

$$= \frac{2n!}{(n!)^2} \frac{((2n+2) - (2n+1))}{n+1}$$

$$= \frac{2n!}{(n!)^2} \left(2 - \frac{2n+1}{n+1} \right)$$

$$= 2 \frac{2n!}{(n!)^2} - \frac{(2n+1)!}{n!(n+1)!}$$

$$= 2 \binom{2n}{n} - \binom{2n+1}{n} .$$

As $\binom{2n}{n}$ and $\binom{2n+1}{n}$ are both integers, this shows that $\frac{1}{n+1}\binom{2n}{n}$ is an integer, as was required to be proved.

Second solution: (due to S. Elnitsky) For $n = 1, 2, \dots$ we have

$$\frac{1}{n+1} \binom{2n}{n} = \frac{2n!}{(n!)^2} \frac{1}{n+1}$$

$$= \frac{2n!}{n!\,(n+1)!}$$

$$= \frac{2n!}{n!\,(n+1)!}\,((n+1)-n)$$

$$= \frac{2n!}{(n!)^2} - \frac{2n!}{(n-1)!\,(n+1)!}$$

$$= \binom{2n}{n} - \binom{2n}{n-1}.$$

As $\binom{2n}{n}$ and $\binom{2n}{n-1}$ are both integers, this shows that $\frac{1}{n+1}\binom{2n}{n}$ is an integer.

62.   Find the sum of the infinite series

$$S = \sum_{n=0}^{\infty} \frac{2^n}{a^{2^n}+1},$$

where $a > 1$.

Solution: We have for $a > 1$

$$\frac{2^n}{a^{2^n}+1} = \frac{2^n(a^{2^n}-1)}{a^{2^{n+1}}-1}$$

$$= \frac{2^n(a^{2^n}+1) - 2^{n+1}}{a^{2^{n+1}}-1}$$

$$= \frac{2^n}{a^{2^n}-1} - \frac{2^{n+1}}{a^{2^{n+1}}-1},$$

so that

$$S = \sum_{n=0}^{\infty} \left(\frac{2^n}{a^{2^n}-1} - \frac{2^{n+1}}{a^{2^{n+1}}-1} \right) = \frac{1}{a-1}.$$

63.   Let k be an integer. Prove that the formal power series

$$\sqrt{1+kx} = 1 + a_1 x + a_2 x^2 + \dots$$

has integral coefficients if and only if $k \equiv 0 \pmod 4$.

Solution: If $k \equiv 1 \pmod 2$ then $a_1 = k/2$ is not an integer and if $k \equiv 2 \pmod 4$ then $a_2 = -k^2/8$ is not an integer. When $k \equiv 0 \pmod 4$, we have for $n = 1, 2, \ldots$

$$
\begin{aligned}
a_n &= \binom{1/2}{n} k^n \\
&= \frac{\frac{1}{2}(\frac{1}{2}-1)(\frac{1}{2}-2)\cdots(\frac{1}{2}-(n-1))}{n!} k^n \\
&= (-1)^{n-1} \frac{1.3.5\cdots(2n-3)}{2^n\, n!} k^n \\
&= (-1)^{n-1} \frac{(2n-2)!}{2^{2n-1}\, n!\, (n-1)!} k^n \\
&= 2(-1)^{n-1} \frac{1}{n} \binom{2n-2}{n-1} \left(\frac{k}{4}\right)^n ,
\end{aligned}
$$

which is an integer since $k/4$ is an integer and $\frac{1}{n}\binom{2n-2}{n-1}$ is an integer by Problem 61.

     64.   Let m be a positive integer. Evaluate the determinant of the $m \times m$ matrix M_m whose (i,j)-th entry is $GCD(i,j)$.

Solution: Let $C_1, \ldots, C_m$ denote the columns of the matrix M_m. We define N_m to the matrix whose columns $D_1, \ldots, D_m$ are given by

$$
\begin{cases}
D_i = C_i, & i = 1, 2, \ldots, m-1 , \\
D_m = \sum_d (-1)^{\tau(d)} C_{m/d} ,
\end{cases}
$$

where the sum is taken over those squarefree integers d which divide m. Clearly, as $D_m = C_m + J$, where J is a linear combination of the C_i, $1 \leq i \leq m-1$, we have

$$
\det M_m = \det N_m .
$$

For $1 \leq i \leq m$, the entry in the i-th row of D_m is (writing (i,j) for $GCD(i,j)$)

$$\sum_{\substack{d|m \\ d \text{ squarefree}}} (-1)^{\tau(d)}(i, m/d) = \prod_{p^\alpha \| m} \sum_{\substack{d|p^\alpha \\ d \text{ squarefree}}} (-1)^{\tau(d)}(i, p^\alpha/d)$$

$$= \prod_{p^\alpha \| m} \left((i, p^\alpha) - (i, p^{\alpha-1}) \right)$$

$$= \begin{cases} \prod_{p^\alpha \| m}(p^\alpha - p^{\alpha-1}) & , \text{ if } i = m , \\ 0 & , \text{ if } 1 \leq i \leq m - 1 , \end{cases}$$

$$= \begin{cases} \phi(m) & , \text{ if } i = m , \\ 0 & , \text{ if } 1 \leq i \leq m - 1 . \end{cases}$$

Hence, expanding the determinant of N_m by its m-th column, we obtain

$$\det N_m = \phi(m) \det N_{m-1}$$

and so

$$\det M_m = \phi(m) \det M_{m-1} .$$

Thus, as $\det M_1 = 1 = \phi(1)$, we find that

$$\det M_m = \phi(m)\phi(m - 1) \cdots \phi(2)\phi(1) .$$

65.   Let l and m be positive integers with l odd and for which there are integers x and y with

$$\begin{cases} l & = x^2 + y^2, \\ m & = x^2 + 8xy + 17y^2. \end{cases}$$

Prove that there do not exist integers u and v with

(65.0) $$\begin{cases} l & = u^2 + v^2, \\ m & = 5u^2 + 16uv + 13v^2. \end{cases}$$

Solution: Suppose there exist integers u and v such that (65.0) holds. Then, we have
$$m = 5l + 8(2uv + v^2) \,,$$
so that $m \equiv 5l \pmod 8$. Hence, we must have
$$x^2 + 8xy + 17y^2 \equiv 5x^2 + 5y^2 \pmod 8 \,,$$
that is
$$4x^2 + 4y^2 \equiv 0 \pmod 8 \,,$$
and so
$$l = x^2 + y^2 \equiv 0 \pmod 2 \,,$$
which contradicts the condition that l is odd.

66.   Let
$$a_n = 1 - \frac{1}{2} + \frac{1}{3} - \ldots + \frac{(-1)^{n-1}}{n} - \ln 2.$$
Prove that $\sum_{n=1}^{\infty} a_n$ converges and determine its sum.

Solution: We have
$$
\begin{aligned}
a_n &= \int_0^1 (1 - x + x^2 - \cdots + (-1)^{n-1} x^{n-1}) \, dx - \int_0^1 \frac{dx}{1+x} \\
&= \int_0^1 \left(\frac{1 + (-1)^{n-1} x^n}{1+x} \right) dx - \int_0^1 \frac{dx}{1+x} \\
&= \int_0^1 \frac{(-1)^{n-1} x^n}{1+x} \, dx \,.
\end{aligned}
$$
Hence, for any integer $N \geq 1$, we have
$$\sum_{n=1}^{N} a_n = \sum_{n=1}^{N} \int_0^1 \frac{(-1)^{n-1} x^n}{1+x} \, dx$$

$$= \int_0^1 \frac{1}{1+x} \sum_{n=1}^N (-1)^{n-1} x^n \, dx$$

$$= \int_0^1 \frac{(x + (-1)^{N+1} x^{N+1})}{(1+x)^2} \, dx$$

$$= \int_0^1 \frac{x}{(1+x)^2} \, dx + (-1)^{N+1} \int_0^1 \frac{x^{N+1}}{(1+x)^2} \, dx \, ,$$

and so

$$\left| \sum_{n=1}^N a_n - \int_0^1 \frac{x}{(1+x)^2} \, dx \right| = \int_0^1 \frac{x^{N+1}}{(1+x)^2} \, dx$$

$$\le \int_0^1 x^{N+1} \, dx$$

$$= \frac{1}{N+2} \, .$$

Letting $N \to \infty$ we see that $\sum_{n=1}^\infty a_n$ converges, and has sum

$$\int_0^1 \frac{x}{(1+x)^2} \, dx = \int_0^1 \left(\frac{1}{1+x} - \frac{1}{(1+x)^2} \right) \, dx = \ln 2 - 1/2 \, .$$

67.   Let $A = \{a_i \mid 0 \le i \le 6\}$ be a sequence of seven integers satisfying

$$0 = a_0 \le a_1 \le \ldots \le a_6 \le 6.$$

For $i = 0, 1, \ldots, 6$ let

$$N_i = \text{ number of } a_j \ (0 \le j \le 6) \text{ such that } a_j = i.$$

Determine all sequences A such that

$$(67.0) \qquad\qquad N_i = a_{6-i} \, , \quad i = 0, 1, \ldots, 6 \, .$$

Solution: Let A be a sequence of the required type satisfying (67.0) and let k denote the number of zeros in A. As $a_0 = 0$ we have $k \geq 1$, and as $k = N_0 = a_6$ we have $k \leq 6$. If $k = 6$ then it follows that $A = \{0,0,0,0,0,0,6\}$, contradicting $N_6 = a_0 = 0$. Hence, we have $1 \leq k = a_6 \leq 5$, and so

$$(67.1) \qquad N_k \geq 1, \quad N_{k+1} = \cdots = N_6 = 0 .$$

Thus, by (67.0) and (67.1), we obtain

$$(67.2) \qquad a_0 = a_1 = \cdots = a_{6-(k+1)} = 0, \quad a_{6-k} \geq 1 ,$$

and so

$$k = N_0 = 6 - (k+1) + 1 ,$$

that is $k = 3$. This proves that A is of the form

$$(67.3) \qquad A = \{0,0,0,a_3,a_4,a_5,3\} ,$$

where

$$(67.4) \qquad 1 \leq a_3 \leq a_4 \leq a_5 \leq 3 .$$

Clearly, we have $0 \leq N_1 \leq 3$. If $N_1 = 0$ then, by (67.0), we have the contradiction $a_5 = N_1 = 0$. If $N_1 = 1$ then, by (67.0), we have $a_5 = 1$, and so (67.4) implies that $a_3 = a_4 = a_5 = 1$, giving the contradiction $N_1 = 3$. If $N_1 = 3$ then $a_3 = a_4 = a_5 = 1$ and so, by (67.0), we obtain the contradiction $a_5 = N_1 = 3$. Hence, we see that $N_1 = 2$ so that $a_3 = a_4 = 1$ and $a_5 = N_1 = 2$. The resulting sequence

$$A = \{0,0,0,1,1,2,3\}$$

satisfies (67.0), and the proof shows that it is the only such sequence to do so.

68. Let G be a finite group with identity e. If G contains elements g and h such that

$$(68.0) \qquad g^5 = e, \quad ghg^{-1} = h^2 ,$$

determine the order of h.

Solution: If $h = e$ then the order of h is 1. Thus we may suppose that $h \neq e$. We have

$$
\begin{aligned}
g^2 h g^{-2} &= g(g h g^{-1})g^{-1} &= g h^2 g^{-1} &= (g h g^{-1})^2 &= h^4 \,, \\
g^3 h g^{-3} &= g(g^2 h g^{-2})g^{-1} &= g h^4 g^{-1} &= (g h g^{-1})^4 &= h^8 \,, \\
g^4 h g^{-4} &= g(g^3 h g^{-3})g^{-1} &= g h^8 g^{-1} &= (g h g^{-1})^8 &= h^{16} \,, \\
g^5 h g^{-5} &= g(g^4 h g^{-4})g^{-1} &= g h^{16} g^{-1} &= (g h g^{-1})^{16} &= h^{32} \,,
\end{aligned}
$$

and so, as $g^5 = e$, we obtain $h = h^{32}$, that is $h^{31} = e$. Thus the order of h is 31 as $h \neq e$ and 31 is prime.

69.   Let a and b be positive integers such that

$$
GCD(a,b) = 1, \quad a \not\equiv b \,(\mathrm{mod}\,2) \,.
$$

If the set S has the following two properties:

        (i) $a, b \in S$,
        (ii) $x, y, z \in S$ implies $x + y + z \in S$,

prove that every integer $> 2ab$ belongs to S.

Solution: Let N be an integer $> 2ab$. As $GCD(a,b) = 1$ there exist integers k and l such that

$$
ak + bl = N \,.
$$

Furthermore, as

$$
\frac{l}{a} - \left(\frac{-k}{b}\right) = \frac{ak + bl}{ab} = \frac{N}{ab} > 2 \,,
$$

there exists an integer t such that

$$
\frac{-k}{b} \leq t < t + 1 \leq \frac{l}{a} \,.
$$

Define integers u and v by

$$u = k + bt, \quad v = l - at,$$

and integers x and y by

$$\begin{cases} x = u, & y = v, & \text{if } u + v \equiv 1 \pmod 2, \\ x = u + b, & y = v - a, & \text{if } u + v \equiv 0 \pmod 2. \end{cases}$$

It is easy to check that

$$N = xa + yb, \quad x \geq 0, \quad y \geq 0, \quad x + y \equiv 1 \pmod 2.$$

We show below that S contains all integers of the form

$$xa + yb, \quad x \geq 0, \quad y \geq 0, \quad x + y \equiv 1 \pmod 2,$$

completing the proof that $N \in S$.

For m an odd positive integer, let P_m be the assertion that $xa + yb \in S$ for all integers x and y satisfying

$$x \geq 0, \quad y \geq 0, \quad x + y \equiv 1 \pmod 2, \quad x + y = m.$$

Clearly P_1 is true as $a, b \in S$ by (i). Assume that P_m is true and consider an integer of the form $Xa + Yb$, where X and Y are integers with

$$X \geq 0, \quad Y \geq 0, \quad X + Y \equiv 1 \pmod 2, \quad X + Y = m + 2.$$

As $m + 2 \geq 3$ at least one of X and Y is ≥ 2. Then, writing $Xa + Yb$ in the form

$$\begin{cases} ((X - 2)a + Yb) + a + a & , \text{ if } X \geq 2, \\ (Xa + (Y - 2)b) + b + b & , \text{ if } Y \geq 2, \end{cases}$$

we see that $Xa + Yb \in S$, by the inductive hypothesis, and so P_{m+2} is true. Hence, by the principle of mathematical induction, P_m is true for all odd positive integers m.

70.   Prove that every integer can be expressed in the form $x^2 + y^2 - 5z^2$, where x, y, z are integers.

Solution: (due to L. Smith)   If m is even, say $m = 2n$, then

$$m = (n - 2)^2 + (2n - 1)^2 - 5(n - 1)^2 \, ,$$

whereas if m is odd, say $m = 2n + 1$, then

$$m = (n + 1)^2 + (2n)^2 - 5n^2 \, .$$

71.   Evaluate the sum of the infinite series

$$\frac{\ln 2}{2} - \frac{\ln 3}{3} + \frac{\ln 4}{4} - \frac{\ln 5}{5} + \cdots \quad .$$

Solution: For $x > 1$ we have

$$\ln x = \int_1^x \frac{dt}{t} < \int_1^x \frac{dt}{\sqrt{t}} = 2\sqrt{x} - 2 < 2\sqrt{x}$$

and

$$-1/2 \le x - [x] - 1/2 < 1/2 \, ,$$

so that for any $a \ge 1$ we have

$$
\begin{aligned}
\int_1^a \left| \frac{(\ln x - 1)}{x^2}(x - [x] - 1/2) \right| \, dx \ &< \ \int_1^a \frac{(2\sqrt{x} + 1)}{x^2} \frac{1}{2} \, dx \\
&< \ \frac{3}{2} \int_1^a \frac{dx}{x^{3/2}} \\
&= \ \frac{3}{2} \left(2 - \frac{2}{\sqrt{a}} \right) \\
&< \ 3 \, .
\end{aligned}
$$

Thus, the integral

$$I = \int_1^\infty \frac{(\ln x - 1)}{x^2}(x - [x] - 1/2) \, dx$$

is absolutely convergent.

Now, one form of the Euler-MacLaurin summation formula asserts that if $f(x)$ has a continuous derivative on $[1, n]$, where n (> 1) is a positive integer, then

$$\sum_{k=1}^{n} f(k) = \frac{1}{2}(f(n) + f(1)) + \int_1^n f(x)\, dx + \int_1^n f'(x)(x - [x] - 1/2)\, dx \ .$$

Taking $f(x) = \ln x/x$, we obtain

$$\sum_{k=1}^{n} \frac{\ln k}{k} = \frac{\ln n}{2n} + \frac{\ln^2 n}{2} + \int_1^n \frac{(1 - \ln x)}{x^2}(x - [x] - 1/2)\, dx \ .$$

Setting

$$E(n) = \sum_{k=1}^{n} \frac{\ln k}{k} - \frac{\ln^2 n}{2} \ ,$$

and letting $n \to \infty$, we see that $\lim_{n\to\infty} E(n)$ exists and has the value $-I$. Thus

$$\lim_{n\to\infty} (E(2n) - E(n))$$

exists and has the value 0. Next, we have the following

$$
\begin{aligned}
\sum_{r=2}^{2n} (-1)^r \frac{\ln r}{r} &= \frac{\ln 2}{2} - \frac{\ln 3}{3} + \frac{\ln 4}{4} - \cdots + \frac{\ln 2n}{2n} \\
&= \left(\frac{\ln 2}{1} + \frac{\ln 4}{2} + \cdots + \frac{\ln 2n}{n}\right) - \left(\frac{\ln 2}{2} + \frac{\ln 3}{3} + \cdots + \frac{\ln 2n}{2n}\right) \\
&= \frac{(\ln 2 + \ln 1)}{1} + \frac{(\ln 2 + \ln 2)}{2} + \cdots + \frac{(\ln 2 + \ln n)}{n} - \sum_{k=1}^{2n} \frac{\ln k}{k} \\
&= \ln 2 \left(1 + \frac{1}{2} + \cdots + \frac{1}{n}\right) + \sum_{k=1}^{n} \frac{\ln k}{k} - \sum_{k=1}^{2n} \frac{\ln k}{k} \\
&= \ln 2 \left(1 + \frac{1}{2} + \cdots + \frac{1}{n}\right) + \left(E(n) + \frac{\ln^2 n}{2}\right) \\
&\qquad - \left(E(2n) + \frac{\ln^2 2n}{2}\right)
\end{aligned}
$$

$$= \ln 2 \left(1 + \frac{1}{2} + \cdots + \frac{1}{n} - \ln n\right) - \frac{\ln^2 2}{2} + (E(n) - E(2n)) .$$

Letting $n \to \infty$, and remembering that

$$\lim_{n \to \infty} \left(\sum_{k=1}^{n} \frac{1}{k} - \ln n\right) = \gamma ,$$

where $\gamma \approx 0.57721$ is Euler's constant, we obtain

$$\sum_{r=2}^{\infty} (-1)^r \frac{\ln r}{r} = \gamma \ln 2 - \frac{1}{2} \ln^2 2 .$$

72.   Determine constants a, b and c such that

$$\sqrt{n} = \sum_{k=0}^{n-1} \sqrt[3]{\sqrt{ak^3 + bk^2 + ck + 1} - \sqrt{ak^3 + bk^2 + ck}} ,$$

for $n = 1, 2, \ldots$.

Solution: For $k = 0, 1, \ldots$, we have

$$
\begin{aligned}
\left(\sqrt{k+1} - \sqrt{k}\right)^3 &= (k+1)\sqrt{k+1} - 3(k+1)\sqrt{k} + 3k\sqrt{k+1} - k\sqrt{k} \\
&= (4k+1)\sqrt{k+1} - (4k+3)\sqrt{k} \\
&= \sqrt{(4k+1)^2(k+1)} - \sqrt{(4k+3)^2 k} \\
&= \sqrt{16k^3 + 24k^2 + 9k + 1} - \sqrt{16k^3 + 24k^2 + 9k}
\end{aligned}
$$

so that

$$\sqrt[3]{\sqrt{16k^3 + 24k^2 + 9k + 1} - \sqrt{16k^3 + 24k^2 + 9k}} = \sqrt{k+1} - \sqrt{k} ,$$

and thus

$$\sum_{k=0}^{n-1} \sqrt[3]{\sqrt{16k^3 + 24k^2 + 9k + 1} - \sqrt{16k^3 + 24k^2 + 9k}}$$

$$= \sum_{k=0}^{n-1} (\sqrt{k+1} - \sqrt{k}) = \sqrt{n}.$$

Hence we may take $a = 16$, $b = 24$, and $c = 9$.

73.   Let n be a positive integer and a, b integers such that

$$GCD(a, b, n) = 1.$$

Prove that there exist integers a_1, b_1 with

$$a_1 \equiv a \, (\text{mod } n), \quad b_1 \equiv b \, (\text{mod } n), \quad GCD(a_1, b_1) = 1.$$

Solution: We choose a_1 to be any nonzero integer such that

(73.1) $a_1 \equiv a \pmod{n}.$

Then we set

$$b_1 = b + rn,$$

where r is the product of those primes which divide a_1 but which do not divide either b or n. If there are no such primes then $r = 1$. Clearly we have

$$b_1 \equiv b \pmod{n}.$$

We now show that

$$GCD(a_1, b_1) = 1.$$

Suppose that $GCD(a_1, b_1) > 1$. Then there exists a prime q which divides both a_1 and b_1. We consider three cases according as

(i) q divides b,
(ii) q does not divide b but divides n,
(iii) q divides neither b nor n.

Case (i): As $q \mid b$, $q \mid b_1$ and $b_1 - b = rn$, we have $q \mid rn$. Now, by (73.1),

$$GCD(a_1, b, n) = GCD(a, b, n) = 1 \ .$$

Since $q \mid a_1$ and $q \mid b$ we see that q does not divide n. Thus we have $q \mid r$, contradicting the definition of r.

Case (ii): This case clearly cannot occur as $b_1 = b + rn$, yet q divides both b_1 and n, but does not divide b.

Case (iii): As $q \mid a_1$ but does not divide b or n, we have $q \mid r$. Since, $q \mid b_1$, $q \mid r$ and $b_1 = b + rn$, we must have $q \mid b$, which is impossible.

This completes the solution.

74.   For $n = 1, 2, \ldots$ let $s(n)$ denote the sum of the digits of 2^n. Thus, for example, as $2^8 = 256$ we have $s(8) = 2 + 5 + 6 = 13$. Determine all positive integers n such that

$$(74.0) \qquad\qquad\qquad s(n) = s(n + 1) \ .$$

Solution: Write

$$2^n = a_m 10^m + a_{m-1} 10^{m-1} + \cdots + a_1 10 + a_0 \ ,$$

where $a_0, a_1, \ldots, a_m$ are integers such that

$$1 \le a_m \le 9 \ ; \quad 0 \le a_k \le 9, \quad 0 \le k \le m - 1 \ ,$$

then

$$2^n \equiv a_m + a_{m-1} + \cdots + a_1 + a_0 \equiv s(n) \pmod{3} \ ,$$

and so

$$s(n + 1) \equiv 2^{n+1} \equiv 2.2^n \equiv 2 \, s(n) \pmod{3} \ .$$

Hence, if $s(n + 1) = s(n)$, we must have

$$s(n) \equiv 0 \pmod{3}, \quad 2^n \equiv 0 \pmod{3} \ ,$$

which is impossible. Thus there are no positive integers satisfying (74.0).

75.   Evaluate the sum of the infinite series

$$S = \sum_{\substack{m,n=1 \\ GCD(m,n)=1}}^{\infty} \frac{1}{mn(m+n)} \, .$$

Solution: We have

$$
\begin{aligned}
\sum_{m,n=1}^{\infty} \frac{1}{mn(m+n)} &= \sum_{m,n=1}^{\infty} \frac{1}{mn} \int_0^1 x^{m+n-1} \, dx \\
&= \int_0^1 \left(\sum_{m=1}^{\infty} \frac{x^m}{m} \right) \left(\sum_{n=1}^{\infty} \frac{x^n}{n} \right) \frac{dx}{x} \\
&= \int_0^1 \frac{\ln^2(1-x)}{x} \, dx \\
&= \int_0^{\infty} \frac{u^2 e^{-u}}{(1-e^{-u})} \, du \qquad (x = 1 - e^{-u}) \\
&= \int_0^{\infty} u^2 \sum_{n=1}^{\infty} e^{-nu} \, du \\
&= \sum_{n=1}^{\infty} \int_0^{\infty} u^2 e^{-nu} \, du \\
&= \sum_{n=1}^{\infty} \frac{2}{n^3} \\
&= 2 \sum_{n=1}^{\infty} \frac{1}{n^3} \, .
\end{aligned}
$$

On the other hand, we have

$$
\sum_{m,n=1}^{\infty} \frac{1}{mn(m+n)} = \sum_{d=1}^{\infty} \sum_{\substack{m,n=1 \\ GCD(m,n)=d}}^{\infty} \frac{1}{mn(m+n)}
$$

$$= \sum_{d=1}^{\infty} \sum_{\substack{q,r=1 \\ GCD(q,r)=1}}^{\infty} \frac{1}{d^3 qr(q+r)}$$

$$= \left(\sum_{d=1}^{\infty} \frac{1}{d^3}\right) \sum_{\substack{q,r=1 \\ GCD(q,r)=1}}^{\infty} \frac{1}{qr(q+r)}$$

$$= S \left(\sum_{d=1}^{\infty} \frac{1}{d^3}\right),$$

so that $S = 2$.

76. A cross-country racer runs a 10-mile race in 50 minutes. Prove that somewhere along the course the racer ran 2 miles in exactly 10 minutes.

Solution: For $0 \leq x \leq 8$ let $T(x)$ denote the time (in minutes) taken by the racer to run between points x and $x + 2$ miles along the course. The function $T(x)$ is continuous on $[0,8]$ and has the property

(76.1) $T(0) + T(2) + T(4) + T(6) + T(8) = 50$.

The equation (76.1) shows that not all of the values $T(0), T(2), T(4), T(6)$ and $T(8)$ are greater than 10 nor are all of them less than 10. Hence, there exist integers r and s with $0 \leq r, s \leq 8$ such that

$$T(r) \leq 10 \leq T(s) .$$

Then, by the intermediate value theorem, there exists a value y, $r \leq y \leq s$, such that $T(y) = 10$, and this proves the assertion.

77. Let AB be a line segment with midpoint O. Let R be a point on AB between A and O. Three semicircles are constructed on the same side of AB as follows: S_1 is the semicircle with centre O and radius $|OA| = |OB|$; S_2 is the semicircle with centre R and radius $|AR|$, meeting RB at C; S_3 is the semicircle with centre S (the midpoint of CB) and radius $|CS| = |SB|$. The common tangent to S_2 and S_3 touches S_2 at P and S_3 at Q. The perpendicular to AB through C meets S_1 at D. Prove that $PCQD$ is a rectangle.

Solution: We give a solution using coordinate geometry. The coordinate system is chosen so that

$$A = (-1, 0), \quad O = (0, 0), \quad B = (1, 0) .$$

Then we have $R = (-a, 0)$, where $0 < a < 1$, and hence

$$C = (1 - 2a, 0), \quad S = (1 - a, 0) .$$

The equations of the three semicircles are given as follows:

$$
\begin{array}{lll}
S_1 & : & x^2 + y^2 = 1 \\
S_2 & : & (x + a)^2 + y^2 = (1 - a)^2 \\
S_3 & : & (x + a - 1)^2 + y^2 = a^2
\end{array}
\quad ,
$$

The perpendicular to AB through C meets S_1 at

$$D = (1 - 2a, 2\sqrt{a - a^2}) .$$

The equation of the common tangent to S_2 and S_3 is

$$x(1 - 2a) + 2y\sqrt{a - a^2} = 1 - 2a + 2a^2 ,$$

and this line touches S_2 at the point

$$P = (2a^2 - 4a + 1, 2(1 - a)\sqrt{a - a^2})$$

and S_3 at the point

$$Q = (1 - 2a^2, 2a\sqrt{a - a^2}) .$$

The slope of PD is

$$\frac{2a\sqrt{a - a^2}}{2a - 2a^2} = \sqrt{\frac{a}{1 - a}} .$$

and the slope of PC is

$$\frac{2(1-a)\sqrt{a-a^2}}{2a^2 - 2a} = -\sqrt{\frac{1-a}{a}} \ .$$

The product of these slopes is -1, showing that PC and PD are perpendicular, that is $\angle CPD = 90°$. Similarly,

$$\angle PDQ = \angle DQC = \angle QCP = 90° \ ,$$

so that $PDQC$ is a rectangle.

78. Determine the inverse of the $n \times n$ matrix

(78.0)
$$S = \begin{bmatrix} 0 & 1 & 1 & \ldots & 1 \\ 1 & 0 & 1 & \ldots & 1 \\ 1 & 1 & 0 & \ldots & 1 \\ \vdots & \vdots & \vdots & \ddots & \vdots \\ 1 & 1 & 1 & \ldots & 0 \end{bmatrix} \ ,$$

where $n \geq 2$.

Solution: Set

$$I = \begin{bmatrix} 1 & 0 & 0 & \ldots & 0 \\ 0 & 1 & 0 & \ldots & 0 \\ 0 & 0 & 1 & \ldots & 0 \\ \vdots & \vdots & \vdots & \ddots & \vdots \\ 0 & 0 & 0 & \ldots & 1 \end{bmatrix} \ , \quad U = \begin{bmatrix} 1 & 1 & 1 & \ldots & 1 \\ 1 & 1 & 1 & \ldots & 1 \\ 1 & 1 & 1 & \ldots & 1 \\ \vdots & \vdots & \vdots & \ddots & \vdots \\ 1 & 1 & 1 & \ldots & 1 \end{bmatrix} \ ,$$

so that

$$S = U - I, \quad U^2 = nU \ .$$

For any real number c, we have

$$\begin{aligned} (U - I)(cU - I) &= cU^2 - (c+1)U + I \\ &= (cn - (c+1))U + I \ . \end{aligned}$$

Thus, if we choose $cn - (c+1) = 0$, that is $c = 1/(n-1)$, we have

$$S^{-1} = (U - I)^{-1} = \frac{1}{n-1}U - I$$

$$= \begin{bmatrix} \frac{2-n}{n-1} & \frac{1}{n-1} & \cdots & \frac{1}{n-1} \\ \frac{1}{n-1} & \frac{2-n}{n-1} & \cdots & \frac{1}{n-1} \\ \vdots & \vdots & \ddots & \vdots \\ \frac{1}{n-1} & \frac{1}{n-1} & \cdots & \frac{2-n}{n-1} \end{bmatrix}.$$

79. Evaluate the sum

$$(79.0) \qquad S(n) = \sum_{k=0}^{n-1} (-1)^k \cos^n(k\pi/n) ,$$

where n is a positive integer.

Solution: Set $\omega = \exp(\pi i/n)$ so that

$$S(n) = \sum_{k=0}^{n-1} (-1)^k \left(\frac{\omega^k + \omega^{-k}}{2} \right)^n .$$

Hence, by the binomial theorem, we obtain

$$\begin{aligned} S(n) &= \frac{1}{2^n} \sum_{k=0}^{n-1} \omega^{kn} \sum_{l=0}^{n} \binom{n}{l} \omega^{k(n-2l)} \\ &= \frac{1}{2^n} \sum_{l=0}^{n} \binom{n}{l} \sum_{k=0}^{n-1} \omega^{k(2n-2l)} \\ &= \frac{1}{2^n} \left(\binom{n}{0} n + \binom{n}{n} n \right) , \end{aligned}$$

that is $S(n) = n/2^{n-1}$.

80. Determine 2×2 matrices B and C with integral entries such that

(80.0)
$$\begin{bmatrix} -1 & 1 \\ 0 & -2 \end{bmatrix} = B^3 + C^3 .$$

Solution: Let

$$A = \begin{bmatrix} -1 & 1 \\ 0 & -2 \end{bmatrix}$$

so that

$$A^2 = \begin{bmatrix} 1 & -3 \\ 0 & 4 \end{bmatrix} ,$$

and thus

$$A^2 + 3A + 2I = 0 ,$$

giving

$$A^3 + 3A^2 + 2A = 0 .$$

Hence, we have

$$(A + I)^3 = A^3 + 3A^2 + 3A + I = A + I ,$$

and so

$$A = (A + I)^3 - I ,$$

and we may take

$$B = A + I = \begin{bmatrix} 0 & 1 \\ 0 & -1 \end{bmatrix} , \quad C = -I = \begin{bmatrix} -1 & 0 \\ 0 & -1 \end{bmatrix} .$$

81. Find two non-congruent similar triangles with sides of integral length having the lengths of two sides of one triangle equal to the lengths of two sides of the other.

Solution: Let the two triangles be ABC and DEF. We suppose that

$$|AB| = a, \quad |AC| = b, \quad |BC| = c,$$
$$|DE| = b, \quad |DF| = c, \quad |EF| = d,$$

and that

(81.1) $$a < b.$$

As $\triangle ABC$ and $\triangle DEF$ are similar, we have

$$\frac{a}{b} = \frac{b}{c} = \frac{c}{d},$$

so that

(81.2) $$c = b^2/a, \quad d = b^3/a^2.$$

From (81.1) we have

(81.3) $$1 < b/a,$$

and from (81.2) and the inequality $c < a + b$ we have

$$\frac{b^2}{a} < a + b,$$

so that

(81.4) $$\frac{b}{a} < \frac{1 + \sqrt{5}}{2} \approx 1.618.$$

To satisfy (81.3) and (81.4) we choose $b/a = 3/2$, say $a = 2t$ and $b = 3t$. Then, by (81.2), we have

$$c = \frac{9t}{2}, \quad d = \frac{27t}{4}.$$

To ensure that c and d are integers we choose $t = 4$ so that

$$a = 8, \quad b = 12, \quad c = 18, \quad d = 27.$$

The triangles with sides $8, 12, 18$ and $12, 18, 27$ respectively, meet the requirements of the problem.

82. Let a, b, c be three real numbers with $a < b < c$. The function $f(x)$ is continuous on $[a, c]$ and differentiable on (a, c). The derivative $f'(x)$ is strictly increasing on (a, c). Prove that

(82.0) $$(c - b)f(a) + (b - a)f(c) > (c - a)f(b).$$

Solution: By the mean-value theorem there exists a real number u such that

$$\frac{f(b) - f(a)}{b - a} = f'(u), \quad a < u < b,$$

and a real number v such that

$$\frac{f(c) - f(b)}{c - b} = f'(v), \quad b < v < c.$$

As $a < u < v < c$ and f' is increasing on (a, c), we have

$$f'(u) < f'(v),$$

and so

$$\frac{f(b) - f(a)}{b - a} < \frac{f(c) - f(b)}{c - b}.$$

Rearranging this inequality gives (82.0).

83. The sequence $\{a_m \mid m = 1, 2, \ldots\}$ is such that $a_m > a_{m+1} > 0$, $m = 1, 2, \ldots$, and $\sum_{m=1}^{\infty} a_m$ converges. Prove that

$$\sum_{m=1}^{\infty} m(a_m - a_{m+1})$$

converges and determine its sum.

Solution: Let $\epsilon > 0$. As $\sum_{m=1}^{\infty} a_m$ is a convergent series of positive terms, there exists a positive integer $N(\epsilon)$ such that

$$(83.1) \qquad 0 < a_{m+1} + a_{m+2} + \cdots < \epsilon/3 \,,$$

for all $m \geq N(\epsilon)$. Let $n \geq 2N(\epsilon)+1$. If n is even, say $n = 2k$, where $k > N(\epsilon)$, from (83.1) we have

$$ka_{2k} < a_{k+1} + a_{k+2} + \cdots + a_{2k} < \epsilon/3 \,,$$

so that

$$na_n = 2ka_{2k} < 2\epsilon/3 < \epsilon \,.$$

If n is odd, say $n = 2k + 1$, where $k \geq N(\epsilon)$, from (83.1) we have

$$ka_{2k+1} < a_{k+2} + a_{k+3} + \cdots + a_{2k+1} < \epsilon/3 \,,$$

so that

$$na_n = 2ka_{2k+1} + a_{2k+1} < 2\epsilon/3 + \epsilon/3 = \epsilon \,.$$

We have shown that

$$0 < na_n < \epsilon, \quad \text{for all } n \geq 2N(\epsilon) + 1 \,,$$

and thus

$$\lim_{n \to \infty} na_n = 0 \,.$$

Next, set

$$S_n = \sum_{k=1}^{n} k(a_k - a_{k+1}), \quad n = 1, 2, \ldots \,.$$

We have

$$\begin{aligned}
S_n &= \sum_{k=1}^{n} ka_k - \sum_{k=1}^{n} ka_{k+1} \\
&= \sum_{k=1}^{n} ka_k - \sum_{k=1}^{n+1} (k-1)a_k
\end{aligned}$$

$$= \sum_{k=1}^{n}(k - (k - 1))a_k - na_{n+1}$$

$$= \sum_{k=1}^{n} a_k - na_{n+1} \ .$$

Letting $n \to \infty$, we see that $\lim_{n \to \infty} S_n$ exists, and has the value $\sum_{k=1}^{\infty} a_k$, as

$$\lim_{n \to \infty} na_{n+1} = \lim_{n \to \infty}((n + 1)a_{n+1} - a_{n+1}) = 0 - 0 = 0 \ .$$

Hence, $\sum_{k=1}^{\infty} k(a_k - a_{k+1})$ converges, and its sum is $\sum_{k=1}^{\infty} a_k$.

84. The continued fraction of $\sqrt{D}$, where D is an odd nonsquare integer > 5, has a period of length one. What is the length of the period of the continued fraction of $\frac{1}{2}(1 + \sqrt{D})$?

Solution: The continued fraction of $\sqrt{D}$ is of the form

$$\sqrt{D} = \left[\, a;\, \overline{b} \,\right] \ ,$$

where a and b are positive integers, so that

$$\sqrt{D} - a = \cfrac{1}{b + \cfrac{1}{b + \cfrac{1}{b + \cdots}}}$$

$$= \frac{1}{b + \sqrt{D} - a} \ ,$$

giving

$$\sqrt{D} = \frac{D + a^2 - ab - 1}{2a - b} \ .$$

As D is not a square, $\sqrt{D}$ is irrational, and we must have

$$b = 2a, \quad D = a^2 + 1 \ .$$

Furthermore, as D is odd and greater than 5, we have $a = 2c$, $c \geq 2$ and $D = 4c^2 + 1$. It is easy to check that

$$\left[\frac{1 + \sqrt{D}}{2}\right] = c,$$

$$\left[\frac{1}{\left(\frac{1+\sqrt{D}}{2}\right) - c}\right] = \left[\frac{2c - 1 + \sqrt{D}}{2c}\right] = 1,$$

$$\left[\frac{1}{\left(\frac{2c-1+\sqrt{D}}{2c}\right) - 1}\right] = \left[\frac{1 + \sqrt{D}}{2c}\right] = 1,$$

$$\left[\frac{1}{\left(\frac{1+\sqrt{D}}{2c}\right) - 1}\right] = \left[\frac{2c - 1 + \sqrt{D}}{2}\right] = 2c - 1,$$

$$\left[\frac{1}{\left(\frac{2c-1+\sqrt{D}}{2}\right) - (2c - 1)}\right] = \left[\frac{2c - 1 + \sqrt{D}}{2c}\right] = 1,$$

so that the continued fraction of $\frac{1}{2}(1 + \sqrt{D})$ is

$$[\, c; \overline{1,\, 1,\, 2c - 1}\,]\,,$$

as $2c - 1 \geq 3$, and its period is of length 3.

85. Let G be a group which has the following two properties:

(85.0)   (i) G has no element of order 2,
                 (ii) $(xy)^2 = (yx)^2$, for all $x, y \in G$.

Prove that G is abelian.

Solution: For $x, y \in G$ we have

$$
\begin{aligned}
x^2 y &= ((xy^{-1})y)^2 y \\
&= (y(xy^{-1}))^2 y \qquad \text{(by (85.0)(ii))} \\
&= (yxy^{-1})(yxy^{-1})y \,,
\end{aligned}
$$

that is

(85.1)
$$x^2 y = y x^2 .$$

Next, we have

$$
\begin{aligned}
x^{-1} y^{-1} x &= x(x^{-1})^2 y^{-1} x \\
&= x y^{-1} (x^{-1})^2 x, \qquad \text{(by (85.1))}
\end{aligned}
$$

that is

(85.2)
$$x^{-1} y^{-1} x = x y^{-1} x^{-1} .$$

Similarly, we have

(85.3)
$$y^{-1} x^{-1} y = y x^{-1} y^{-1} .$$

Then we obtain

$$
\begin{aligned}
(x y x^{-1} y^{-1})^2 &= x y (x^{-1} y^{-1} x) y x^{-1} y^{-1} \\
&= x y (x y^{-1} x^{-1}) y x^{-1} y^{-1} \qquad \text{(by (85.2))} \\
&= x y x (y^{-1} x^{-1} y) x^{-1} y^{-1} \\
&= x y x (y x^{-1} y^{-1}) x^{-1} y^{-1} \qquad \text{(by (85.3))} \\
&= (x y)^2 (x^{-1} y^{-1})^2 \\
&= (x y)^2 (y x)^{-2} \\
&= (y x)^2 (y x)^{-2} \qquad \text{(by (85.0)(ii))} \\
&= 1 ,
\end{aligned}
$$

and thus, as G has no elements of order 2, we have

$$x y x^{-1} y^{-1} = 1 ,$$

that is $xy = yx$, proving that G is abelian.

86.   Let $A = [a_{ij}]$ be an $n \times n$ real symmetric matrix whose entries satisfy

(86.0)
$$a_{ii} = 1 , \qquad \sum_{j=1}^{n} |a_{ij}| \leq 2 ,$$

for all $i = 1, 2, \ldots, n$. Prove that $0 \leq \det A \leq 1$.

Solution: Let λ denote one of the eigenvalues of A and let $\underline{x}$ ($\neq \underline{0}$) be an eigenvector of A corresponding to λ, so that

(86.1) $$A\underline{x} = \lambda \underline{x} .$$

Set $\underline{x} = (x_1, \ldots, x_n)^t$ and choose i, $1 \leq i \leq n$, so that

$$|x_i| = \max_{1 \leq j \leq n} |x_j| \neq 0 .$$

From the i-th row of (86.1), we obtain

$$\sum_{j=1}^{n} a_{ij} x_j = \lambda x_i ,$$

so that

$$(\lambda - 1)x_i = \sum_{\substack{j=1 \\ j \neq i}}^{n} a_{ij} x_j ,$$

and thus

$$
\begin{aligned}
|\lambda - 1||x_i| &= |\sum_{\substack{j=1 \\ j \neq i}}^{n} a_{ij} x_j| \\
&\leq \sum_{\substack{j=1 \\ j \neq i}}^{n} |a_{ij}||x_j| \\
&\leq |x_i| \sum_{\substack{j=1 \\ j \neq i}}^{n} |a_{ij}| \\
&\leq |x_i| ,
\end{aligned}
$$

showing that

(86.2) $$|\lambda - 1| \leq 1 .$$

Since A is a real symmetric matrix, λ is real and from (86.2) we see that

(86.3) $$0 \le \lambda \le 2 \ .$$

Let $\lambda_1, \ldots, \lambda_n$ denote the n eigenvalues of A. Each λ_j is nonnegative by (86.3). Thus we have

$$
\begin{aligned}
0 \le \det A &= \lambda_1 \lambda_2 \cdots \lambda_n \\
&\le \left(\frac{1}{n} (\lambda_1 + \lambda_2 + \cdots + \lambda_n) \right)^n \\
&= \left(\frac{1}{n} \operatorname{trace} A \right)^n \\
&= \left(\frac{1}{n} n \right)^n \\
&= 1 \ .
\end{aligned}
$$

87.   Let R be a finite ring containing an element r which is not a divisor of zero. Prove that R must have a multiplicative identity.

Solution: As R is a finite ring there exist integers m and n such that

(87.1) $$r^m = r^n, \quad 1 \le m < n \ .$$

We wish to show that

(87.2) $$r = r^k \ ,$$

for some integer $k \ge 2$. If $m = 1$ we may take $k = n$. If $m \ge 2$, from (87.1), we have

$$r(r^{m-1} - r^{n-1}) = 0 \ .$$

As r is not a divisor of zero, we must have

(87.3) $$r^{m-1} - r^{n-1} = 0 \ .$$

If $m = 2$ we may take $k = n - 1$ (≥ 2). If $m \ge 3$, from (87.3) we have

$$r(r^{m-2} - r^{n-2}) = 0 \ .$$

As r is not a divisor of zero, we must have

$$r^{m-2} - r^{n-2} = 0 .$$

If $m = 3$ we may take $k = n - 2 \, (\geq 2)$. Continuing in this way, we see that (87.2) holds with $k = n - m + 1 \, (\geq 2)$. For any $x \in R$ we have from (87.2)

$$xr = xr^k$$

and so

$$(x - xr^{k-1})r = 0 .$$

As r is not a divisor of zero, we see that

(87.4) $$x = xr^{k-1} .$$

Similarly, we have

(87.5) $$x = r^{k-1}x .$$

From (87.4) and (87.5) we see that r^{k-1} is a multiplicative identity for R.

88. Set $J_n = \{1, 2, \ldots, n\}$. For each non-empty subset S of J_n define

$$w(S) = \max_{s \in S} S - \min_{s \in S} S .$$

Determine the average of $w(S)$ over all non-empty subsets S of J_n.

Solution: For $1 \leq k \leq l \leq n$ let $S(k,l)$ denote the set of subsets of J_n with

$$\min_{s \in S} S = k, \quad \max_{s \in S} S = l .$$

We have, for all $S \in S(k,l)$,

$$w(S) = l - k ,$$

and

$$|S(k,l)| = \begin{cases} 1 , & \text{if } k = l , \\ 2^{l-k-1} , & \text{if } k < l , \end{cases}$$

Then we have

$$
\begin{aligned}
\sum_{\phi \neq S \subseteq J_n} w(S) &= \sum_{1 \leq k \leq l \leq n} \sum_{S \in S(k,l)} w(S) \\
&= \sum_{1 \leq k < l \leq n} (l - k)|S(k,l)| \\
&= \sum_{1 \leq k \leq l \leq n} (l - k)2^{l-k-1} \\
&= \sum_{k=1}^{n-1} 2^{-k-1} \sum_{l=k+1}^{n} l 2^l - \sum_{k=1}^{n-1} k 2^{-k-1} \sum_{l=k+1}^{n} 2^l \\
&= \sum_{k=1}^{n-1} 2^{-k-1} \left((n-1)2^{n+1} - (k-1)2^{k+1} \right) \\
&\qquad - \sum_{k=1}^{n-1} k 2^{-k-1}(2^{n+1} - 2^{k+1}) \\
&= (n-1)2^n \sum_{k=1}^{n-1} 2^{-k} - \sum_{k=1}^{n-1}(k-1) \\
&\qquad - 2^n \sum_{k=1}^{n-1} k 2^{-k} + \sum_{k=1}^{n-1} k \\
&= (n-1)2^n \left(1 - \frac{1}{2^{n-1}} \right) \\
&\qquad - 2^n \left(2 - \frac{(n+1)}{2^{n-1}} \right) + n - 1 \\
&= (n-1)2^n - 2(n-1) - 2^{n+1} + 2(n+1) + n - 1 \\
&= (n-3)2^n + (n+3) \,,
\end{aligned}
$$

so that the required average is

$$
\frac{(n-3)2^n + (n+3)}{2^n - 1}, \quad n = 1, 2, \ldots .
$$

89.   Prove that the number of odd binomial coefficients in each row of Pascal's triangle is a power of 2.

Solution: The entries in the n-th row of Pascal's triangle are the coefficients of the powers of x in the expansion of $(1 + x)^n$. We write n in binary notation

$$(89.1) \qquad\qquad n = 2^{a_1} + 2^{a_2} + \cdots + 2^{a_k} \, ,$$

where $a_1, \ldots, a_k$ are integers such that

$$(89.2) \qquad\qquad a_1 > a_2 > \cdots > a_k \geq 0 \, .$$

Now

$$\begin{aligned}
(1 + x)^2 &= 1 + 2x + x^2 &\equiv 1 + x^2 &\quad (\mathrm{mod}\ 2), \\
(1 + x)^4 &\equiv (1 + x^2)^2 &\equiv 1 + x^4 &\quad (\mathrm{mod}\ 2), \\
(1 + x)^8 &\equiv (1 + x^4)^2 &\equiv 1 + x^8 &\quad (\mathrm{mod}\ 2),
\end{aligned}$$

and so generally for any nonnegative integer a we have

$$(1 + x)^{2^a} \equiv 1 + x^{2^a} \quad (\mathrm{mod}\ 2) \, .$$

Thus, we have

$$\begin{aligned}
(1 + x)^n &= (1 + x)^{2^{a_1} + 2^{a_2} + \cdots + 2^{a_k}} \\
&= (1 + x)^{2^{a_1}} (1 + x)^{2^{a_2}} \cdots (1 + x)^{2^{a_k}} \\
&\equiv (1 + x^{2^{a_1}})(1 + x^{2^{a_2}}) \cdots (1 + x^{2^{a_k}}) \quad (\mathrm{mod}\ 2) \\
&\equiv 1 + (x^{2^{a_1}} + x^{2^{a_2}} + \cdots + x^{2^{a_k}}) \\
&\qquad + \left(x^{2^{a_1} + 2^{a_2}} + \cdots + x^{2^{a_{k-1}} + 2^{a_k}} \right) \\
&\qquad + \cdots \\
&\qquad + x^{2^{a_1} + 2^{a_2} + \cdots + 2^{a_k}} \quad (\mathrm{mod}\ 2) \, ,
\end{aligned}$$

and the number of odd coefficients is

$$1 + k + \binom{k}{2} + \cdots + \binom{k}{k} = 2^k \, .$$

90. From the $n \times n$ array

$$\begin{bmatrix} 1 & 2 & 3 & \cdots & n \\ n+1 & n+2 & n+3 & \cdots & 2n \\ 2n+1 & 2n+2 & 2n+3 & \cdots & 3n \\ \vdots & \vdots & \vdots & \ddots & \vdots \\ (n-1)n+1 & (n-1)n+2 & (n-1)n+3 & \cdots & n^2 \end{bmatrix}$$

a number x_1 is selected. The row and column containing x_1 are then deleted. From the resulting array a number x_2 is selected, and its row and column deleted as before. The selection is continued until only one number x_n remains available for selection. Determine the sum $x_1 + x_2 + \cdots + x_n$.

Solution: Suppose that x_i, $1 \le i \le n$, belongs to the r_i-th row and the s_i-th column of the array. Then

$$x_i = (r_i - 1)n + s_i, \quad 1 \le i \le n,$$

and so

$$\sum_{i=1}^{n} x_i = n \sum_{i=1}^{n} r_i - n^2 + \sum_{i=1}^{n} s_i.$$

Now $\{r_1, \ldots, r_n\}$ and $\{s_1, \ldots, s_n\}$ are permutations of $\{1, 2, \ldots, n\}$ and so

$$\sum_{i=1}^{n} r_i = \sum_{i=1}^{n} s_i = \sum_{i=1}^{n} i = \frac{n(n+1)}{2}.$$

Thus

$$\sum_{i=1}^{n} x_i = \frac{n^2(n+1)}{2} - n^2 + \frac{n(n+1)}{2} = \frac{n(n^2+1)}{2}.$$

91. Suppose that p X's and q O's are placed on the circumference of a circle. The number of occurrences of two adjacent X's is a and the number of occurrences of two adjacent O's is b. Determine $a - b$ in terms of p and q.

Solution: Let

$$N_{xx}, \quad N_{xo}, \quad N_{ox}, \quad N_{oo}$$

denote the number of occurrences of XX, XO, OX, OO, respectively. Then clearly we have

$$\begin{cases} N_{xx} = a\,, \\ N_{oo} = b\,, \\ N_{xo} + N_{xx} = p\,, \\ N_{oo} + N_{ox} = q\,, \end{cases}$$

so that

$$\begin{aligned} a - b &= N_{xx} - N_{oo} \\ &= (N_{xx} + N_{xo}) - (N_{oo} + N_{ox}) + (N_{ox} - N_{xo}) \\ &= p - q + (N_{ox} - N_{xo})\,. \end{aligned}$$

Finally, we show that $N_{ox} = N_{xo}$, which gives the result

$$a - b = p - q\,.$$

To see that $N_{ox} = N_{xo}$ we consider the values of a function S as we make one clockwise tour of the circumference of the circle, starting and finishing at the same point. Initially, we let $S = 0$. Then, as we tour the circle, the value of S is changed as follows as we pass from each X or O to the next X or O:

$$\text{new value of } S = \text{old value of } S + \epsilon,$$

where

$$\epsilon = \begin{cases} 1 & \text{, in going form O to X}\,, \\ 0 & \text{, in going from X to X or O to O}\,, \\ -1 & \text{, in going from X to O}\,. \end{cases}$$

Clearly, the value of S at the end of the tour is $N_{ox} - N_{xo}$. However, S must be 0 at the end as we have returned to the starting point. This completes the proof of $N_{ox} = N_{xo}$, and the solution.

92. In the triangular array

$$
\begin{array}{ccccccccc}
 & & & & 1 & & & & \\
 & & & 1 & 1 & 1 & & & \\
 & & 1 & 2 & 3 & 2 & 1 & & \\
 & 1 & 3 & 6 & 7 & 6 & 3 & 1 & \\
1 & 4 & 10 & 16 & 19 & 16 & 10 & 4 & 1
\end{array}
$$

(92.0)

$$\cdot \quad \cdot \quad \cdot$$

every entry (except the top 1) is the sum of the entry a immediately above it, and the entries b and c immediately to the left and right of a. Absence of an entry indicates zero. Prove that every row after the second row contains an entry which is even.

Solution: The first eight rows of the triangular array taken modulo 2 are given in (92.1).

(92.1)

$$
\begin{array}{ccccccccccccccc}
 & & & & & & & 1 & & & & & & & \\
 & & & & & & 1 & 1 & 1 & & & & & & \\
 & & & & & 1 & 0 & 1 & 0 & 1 & & & & & \\
 & & & & 1 & 1 & 0 & 1 & 0 & 1 & 1 & & & & \\
 & & & 1 & 0 & 0 & 0 & 1 & 0 & 0 & 0 & 1 & & & \\
 & & 1 & 1 & 1 & 0 & 1 & 1 & 1 & 0 & 1 & 1 & 1 & & \\
 & 1 & 0 & 1 & 0 & 0 & 0 & 1 & 0 & 0 & 0 & 1 & 0 & 1 & \\
1 & 1 & 0 & 1 & 1 & 0 & 1 & 1 & 1 & 0 & 1 & 1 & 0 & 1 & 1
\end{array}
$$

The first four entries in the fourth row of (92.1) are 1 1 0 1, which are exactly the same as the first four entries in the eighth row. Thus the pattern

(92.2)

$$
\begin{array}{cccc}
1 & 1 & 0 & 1 \\
1 & 0 & 0 & 0 \\
1 & 1 & 1 & 0 \\
1 & 0 & 1 & 0
\end{array}
$$

repeats itself down the left-hand edge of the array. As each row of (92.2) contains at least one zero, every row from the fourth on down contains an even number. This completes the proof, as the third row contains an even number.

93. A sequence of n real numbers $x_1, \ldots, x_n$ satisfies

(93.0)
$$\begin{cases} x_1 = 0, \\ |x_i| = |x_{i-1} + c|, \quad 2 \leq i \leq n, \end{cases}$$

where c is a positive real number. Determine a lower bound for the average of $x_1, \ldots, x_n$ as a function of c only.

Solution: Let x_{n+1} be any real number such that

$$|x_{n+1}| = |x_n + c|.$$

Then, we have

$$\begin{aligned} \sum_{i=1}^{n+1} x_i^2 &= \sum_{i=2}^{n+1} |x_i|^2 = \sum_{i=2}^{n+1} |x_{i-1} + c|^2 \\ &= \sum_{i=2}^{n+1} (x_{i-1} + c)^2 \\ &= \sum_{i=2}^{n+1} x_{i-1}^2 + 2c \sum_{i=2}^{n+1} x_{i-1} + c^2 n \\ &= \sum_{i=1}^{n} x_i^2 + 2c \sum_{i=1}^{n} x_i + c^2 n, \end{aligned}$$

so that

$$0 \leq x_{n+1}^2 = 2c \sum_{i=1}^{n} x_i + c^2 n,$$

and thus (as $c > 0$)

$$\frac{1}{n} \sum_{i=1}^{n} x_i \geq -\frac{c}{2}.$$

94.   Prove that the polynomial

(94.0) $$f(x) = x^n + x^3 + x^2 + x + 5$$

is irreducible over **Z** for $n \geq 4$.

Solution: Suppose $f(x)$ is reducible over **Z**. Then there exist monic polynomials $g(x)$ and $h(x)$ with integral coefficients such that

(94.1) $$f(x) = g(x)h(x), \quad \deg g \geq 1, \quad \deg h \geq 1 .$$

Thus, we have

$$5 = f(0) = g(0)h(0) ,$$

and, as $g(0)$, $h(0)$ are integers and 5 is prime, we have without loss of generality

$$g(0) = \pm 1, \quad h(0) = \pm 5 .$$

Let

$$g(x) = \prod_{j=1}^{r} (x - \beta_j)$$

be the factorization of $g(x)$ over **C**. Then, we have

$$1 = |g(0)| = \prod_{j=1}^{r} |\beta_j| ,$$

and so at least one of the $|\beta_j|$ is less than or equal to 1, say

$$|\beta_l| \leq 1, \quad 1 \leq l \leq r .$$

Hence

$$
\begin{aligned}
|f(\beta_l)| &= |\beta_l^n + \beta_l^3 + \beta_l^2 + \beta_l + 5| \\
&\geq 5 - |\beta_l| - |\beta_l|^2 - |\beta_l|^3 - |\beta_l|^n \\
&\geq 5 - 1 - 1 - 1 - 1 \\
&= 1 ,
\end{aligned}
$$

which contradicts

$$f(\beta_l) = g(\beta_l)h(\beta_l) = 0 \ h(\beta_l) = 0 \ .$$

This proves that $f(x)$ is irreducible over $\mathbf{Z}$.

95. Let $a_1, \ldots, a_n$ be $n \ (\geq 4)$ distinct real numbers. Determine the general solution of the system of $n - 2$ linear equations

$$(95.0) \quad \begin{cases} x_1 + \quad x_2 + \cdots + \quad x_n & = 0, \\ a_1 x_1 + a_2 x_2 + \cdots + a_n x_n & = 0, \\ a_1^2 x_1 + a_2^2 x_2 + \cdots + a_n^2 x_n & = 0, \\ \quad \vdots \\ a_1^{n-3} x_1 + a_2^{n-3} x_2 + \cdots + a_n^{n-3} x_n & = 0, \end{cases}$$

in the n unknowns $x_1, \ldots, x_n$.

Solution: Set

$$f(x) = (x - a_1)(x - a_2) \cdots (x - a_n) \ .$$

For $k = 0, 1, \ldots, n - 1$ the partial fraction expansion of $x^k/f(x)$ is

$$(95.1) \quad \frac{x^k}{f(x)} = \sum_{i=1}^{n} \frac{a_i^k/f'(a_i)}{x - a_i} \ .$$

Multiplying both sides of (95.1) by $f(x)$, and equating coefficients of x^{n-1}, we obtain

$$(95.2) \quad \sum_{i=1}^{n} \frac{a_i^k}{f'(a_i)} = \begin{cases} 0 & , k = 0, 1, \ldots, n - 2 \ , \\ 1 & , k = n - 1 \end{cases}$$

This shows that

$$\underline{u} = \left(\frac{1}{f'(a_1)}, \ldots, \frac{1}{f'(a_n)} \right)$$

and

$$\underline{v} = \left(\frac{a_1}{f'(a_1)}, \dots, \frac{a_n}{f'(a_n)} \right)$$

are two solutions of (95.0). These two solutions are linearly independent, for otherwise there would exist real numbers s and t (not both zero) such that

$$s\underline{u} + t\underline{v} = (0, \dots, 0),$$

that is

(95.3) $\qquad\qquad s + ta_i = 0, \quad i = 1, 2, \dots, n.$

If $t = 0$ then from (95.3) we have $s = 0$, which is a contradiction. Thus, $t \neq 0$ and (95.3) gives

$$a_i = -\frac{s}{t}, \quad i = 1, 2, \dots, n,$$

which contradicts the fact that the a_i are distinct. Thus the solutions $\underline{u}$ and $\underline{v}$ are linearly independent.

Next, as the a_i are distinct, the Vandermonde determinant

$$\begin{vmatrix} 1 & 1 & \cdots & 1 \\ a_1 & a_2 & \cdots & a_{n-2} \\ a_1^2 & a_2^2 & \cdots & a_{n-2}^2 \\ \vdots & \vdots & \ddots & \vdots \\ a_1^{n-3} & a_2^{n-3} & \cdots & a_{n-2}^{n-3} \end{vmatrix}$$

does not vanish, and so the rank of the coefficient matrix of (95.0) is $n - 2$. Thus all solutions of (95.0) are given as linear combinations of any two linearly independent solutions. Hence all solutions of (95.0) are given by

$$\begin{aligned} (x_1, \dots, x_n) &= \alpha\underline{u} + \beta\underline{v} \\ &= \left(\frac{\alpha + \beta a_1}{f'(a_1)}, \dots, \frac{\alpha + \beta a_n}{f'(a_n)} \right), \end{aligned}$$

for real numbers α and β.

96. Evaluate the sum

$$S(N) = \sum_{\substack{1 \le m < n \le N \\ m+n > N \\ GCD(m,n)=1}} \frac{1}{mn}, \quad N = 2, 3, \ldots.$$

Solution: For $N \ge 3$ we have

$$
\begin{aligned}
S(N) &= \sum_{\substack{1 \le m < n \le N-1 \\ m+n > N \\ GCD(m,n)=1}} \frac{1}{mn} + \sum_{\substack{1 \le m < n = N \\ m+n > N \\ GCD(m,n)=1}} \frac{1}{mn} \\[2ex]
&= \sum_{\substack{1 \le m < n \le N-1 \\ m+n > N-1 \\ GCD(m,n)=1}} \frac{1}{mn} - \sum_{\substack{1 \le m < n \le N-1 \\ m+n = N \\ GCD(m,n)=1}} \frac{1}{mn} + \sum_{\substack{1 \le m < N \\ GCD(m,N)=1}} \frac{1}{mN} \\[2ex]
&= S(N-1) - \sum_{\substack{1 \le m < N/2 \\ GCD(m,N)=1}} \frac{1}{m(N-m)} + \frac{1}{N} \sum_{\substack{1 \le m < N \\ GCD(m,N)=1}} \frac{1}{m} \\[2ex]
&= S(N-1) - \frac{1}{N} \sum_{\substack{1 \le m < N/2 \\ GCD(m,N)=1}} \frac{1}{m} - \frac{1}{N} \sum_{\substack{1 \le m < N/2 \\ GCD(m,N)=1}} \frac{1}{N-m} \\[2ex]
&\qquad + \frac{1}{N} \sum_{\substack{1 \le m < N \\ GCD(m,N)=1}} \frac{1}{m}
\end{aligned}
$$

$$= S(N-1) \quad - \quad \frac{1}{N} \sum_{\substack{1 \le m < N \\ GCD(m,N)=1}} \frac{1}{m} \quad + \quad \frac{1}{N} \sum_{\substack{1 \le m < N \\ GCD(m,N)=1}} \frac{1}{m}$$

$$= S(N-1) ,$$

remembering that $GCD(N/2, N) > 1$ for even $N \ (\ge 4)$. Thus, we have

$$S(N) = S(N-1) = S(N-2) = \cdots = S(2) = 1/2 .$$

97.   Evaluate the limit

(97.0)
$$L = \lim_{n \to \infty} \frac{1}{n} \sum_{j=1}^{n} \sum_{k=1}^{n} \frac{j}{j^2 + k^2} .$$

Solution: Partition the unit square $[0, 1] \times [0, 1]$ into n^2 subsquares by the partition points

$$\{ (j/n, k/n) : \ 0 \le j, k \le n \} .$$

Then a Riemann sum of the function $x/(x^2 + y^2)$ for this partition is

$$\sum_{1 \le j,k \le n} \frac{j/n}{(j/n)^2 + (k/n)^2} \frac{1}{n^2} \quad = \quad \frac{1}{n} \sum_{1 \le j,k \le n} \frac{j}{j^2 + k^2} ,$$

and also

$$\lim_{n \to \infty} \sum_{1 \le j,k \le n} \frac{j/n}{(j/n)^2 + (k/n)^2} \frac{1}{n^2} \quad = \quad \int\int_{[0,1] \times [0,1]} \frac{x}{x^2 + y^2} \, dx \, dy ,$$

so that (97.0) becomes

$$L \quad = \quad \int_0^1 \int_0^1 \frac{x}{x^2 + y^2} \, dx \, dy$$

$$= \int_{\theta=0}^{\pi/4} \int_{r=0}^{\sec\theta} \cos\theta \; dr \; d\theta \;\; + \;\; \int_{\theta=\pi/4}^{\pi/2} \int_{r=0}^{\csc\theta} \cos\theta \; dr \; d\theta$$

$$= \int_{0}^{\pi/4} d\theta \;\; + \;\; \int_{\pi/4}^{\pi/2} \cot\theta \; d\theta$$

$$= \pi/4 + [\, \ln\sin\theta \,]_{\pi/4}^{\pi/2}$$

$$= \pi/4 - \ln(1/\sqrt{2}) \;,$$

that is $L = \pi/4 + (\ln 2)/2$.

98.   Prove that

(98.0) $$\tan\frac{3\pi}{11} + 4\sin\frac{2\pi}{11} = \sqrt{11} \;.$$

Solution: For convenience we let $p = \pi/11$, and set

$$c = \cos p, \quad s = \sin p \;.$$

Then, we have $c + is = e^{pi}$ and so $(c + is)^{11} = -1$, that is

$$c^{11} + 11c^{10}si - 55c^9 s^2 - 165 c^8 s^3 i + 330 c^7 s^4 + 462 c^6 s^5 i$$
$$-462 c^5 s^6 - 330 c^4 s^7 i + 165 c^3 s^8 + 55 c^2 s^9 i - 11 c s^{10} - s^{11} i = -1 \;.$$

Equating imaginary parts, we obtain

(98.1) $$11 c^{10} s - 165 c^8 s^3 + 462 c^6 s^5 - 330 c^4 s^7 + 55 c^2 s^9 - s^{11} = 0 \;.$$

From (98.1), as $s \neq 0$, we have

(98.2) $$11 c^{10} - 165 c^8 s^2 + 462 c^6 s^4 - 330 c^4 s^6 + 55 c^2 s^8 - s^{10} = 0 \;.$$

Next, as

(98.3) $$c^2 = 1 - s^2 \;,$$

the equation (98.2) becomes

(98.4) $$11 - 220 s^2 + 1232 s^4 - 2816 s^6 + 2816 s^8 - 1024 s^{10} = 0 \;,$$

and thus

$$(11s - 44s^3 + 32s^5)^2 - 11c^2(1 - 4s^2)^2$$
$$= 121s^2 - 968s^4 + 2640s^6 - 2816s^8 + 1024s^{10}$$
$$-11(1 - s^2)(1 - 8s^2 + s^4)$$
$$= -11 + 220s^2 - 1232s^4 + 2816s^6 - 2816s^8 + 1024s^{10}$$
$$= 0 \, ,$$

by (98.4). This proves that

(98.5)
$$\frac{11s - 44s^3 + 32s^5}{c(1 - 4s^2)} = \pm\sqrt{11} \, .$$

Next, we have

$$\tan 3p + 4\sin 2p \quad = \quad \frac{3\tan p - \tan^3 p}{1 - 3\tan^2 p} + 8\sin p \, \cos p$$
$$= \quad \frac{3sc^2 - s^3}{c^3 - 3s^2 c} + 8s\,c \, ,$$

that is, using (98.3),

(98.6)
$$\tan 3p + 4\sin 2p = \frac{11s - 44s^3 + 32s^5}{c(1 - 4s^2)} \, .$$

Then, from (98.5) and (98.6), we obtain

$$\tan 3p + 4\sin 2p = \pm\sqrt{11} \, .$$

As $\tan 3p > 0$, $\sin 2p > 0$, we must have

$$\tan \frac{3\pi}{11} + 4\sin \frac{2\pi}{11} = \sqrt{11} \, ,$$

as required.

99.   For $n = 1, 2, \ldots$ let

$$c_n = 1 + \frac{1}{2} + \frac{1}{3} + \cdots + \frac{1}{n} \, .$$

Evaluate the sum

$$S = \sum_{n=1}^{\infty} \frac{c_n}{n(n+1)} \ .$$

Solution: Let k be a positive integer. We have

$$\sum_{n=1}^{k} \frac{c_n}{n(n+1)} = \sum_{n=1}^{k} \left(\frac{c_n}{n} - \frac{c_n}{n+1} \right)$$

$$= \sum_{n=1}^{k} \frac{c_n}{n} - \sum_{n=2}^{k+1} \frac{c_{n-1}}{n}$$

$$= c_1 + \sum_{n=2}^{k} \frac{(c_n - c_{n-1})}{n} - \frac{c_k}{k+1}$$

$$= 1 + \sum_{n=2}^{k} \frac{1}{n^2} - \frac{c_k}{k+1}$$

$$= \sum_{n=1}^{k} \frac{1}{n^2} - \frac{c_k}{k+1}$$

$$= \sum_{n=1}^{k} \frac{1}{n^2} - \frac{(c_k - \ln k)}{k+1} - \frac{\ln k}{k+1} \ .$$

Letting $n \to \infty$, and using the fact that

$$\lim_{k \to \infty} (c_k - \ln k)$$

exists, and also

$$\lim_{k \to \infty} \frac{\ln k}{k+1} = 0 \ ,$$

we find that

$$S = \sum_{n=1}^{\infty} \frac{1}{n^2} = \frac{\pi^2}{6} \ .$$

100. For $x > 1$ determine the sum of the infinite series

$$\frac{x}{x+1} + \frac{x^2}{(x+1)(x^2+1)} + \frac{x^4}{(x+1)(x^2+1)(x^4+1)} + \cdots .$$

Solution: For n a positive integer, set

$$S_n(x) = \frac{x}{x+1} + \frac{x^2}{(x+1)(x^2+1)} + \cdots + \frac{x^{2^n}}{(x+1)(x^2+1)\cdots(x^{2^n}+1)} ,$$

so that

$$\begin{aligned}
\frac{S_n(x)}{x-1} &= \frac{x}{x^2-1} + \frac{x^2}{x^4-1} + \cdots + \frac{x^{2^n}}{x^{2^{n+1}}-1} \\
&= \left(\frac{1}{x-1} - \frac{1}{x^2-1} \right) + \left(\frac{1}{x^2-1} - \frac{1}{x^4-1} \right) \\
&\qquad + \cdots + \left(\frac{1}{x^{2^n}-1} - \frac{1}{x^{2^{n+1}}-1} \right) \\
&= \frac{1}{x-1} - \frac{1}{x^{2^{n+1}}-1} .
\end{aligned}$$

Thus, as $x > 1$, we have

$$\lim_{n\to\infty} \frac{S_n(x)}{x-1} = \frac{1}{x-1}$$

giving

$$\frac{x}{x+1} + \frac{x^2}{(x+1)(x^2+1)} + \cdots = \lim_{n\to\infty} S_n(x) = 1 .$$

THE SOURCES

Problem

01: Gauss, see *Werke*, Vol 2, Göttingen (1876), pp.11-45, showed that

$$\omega^{r_1} + \omega^{r_2} + \cdots + \omega^{r_{(p-1)/2}} = (-1 + i^{\left(\frac{p-1}{2}\right)^2}\sqrt{p})/2 \,,$$
$$\omega^{n_1} + \omega^{n_2} + \cdots + \omega^{n_{(p-1)/2}} = (-1 - i^{\left(\frac{p-1}{2}\right)^2}\sqrt{p})/2 \,.$$

04: This result is implicit in the work of Gauss, see *Werke*, Vol 2, Göttingen (1876), p.292.

05: The more general equation $y^2 = x^3 + ((4b-1)^3 - 4a^2)$, where a has no prime factors $\equiv 3 \pmod 4$, is treated in L.J. Mordell, *Diophantine Equations*, Academic Press (1969), pp.238-239.

09: This problem was suggested by Problem 97 of *The Green Book*. It also appears as Problem E2115 in American Mathematical Monthly 75 (1968), p.897 with a solution by G.V. McWilliams in American Mathematical Monthly 76 (1969), p.828.

10: This problem is due to Professor Charles A. Nicol of the University of South Carolina.

11: Another solution to this problem is given in Crux Mathematicorum 14 (1988), pp.19-20.

14: The more general equation $dV^2 - 2eVW - dW^2 = 1$ is treated in K. Hardy and K.S. Williams, *On the solvability of the diophantine equation $dV^2 - 2eVW - dW^2 = 1$*, Pacific Journal of Mathematics 124 (1986), pp.145-158.

17: This generalizes the well-known result that the sequence $1, 2, \ldots, 10$ contains a pair of consecutive quadratic residues modulo a prime ≥ 11. The required pair can be taken to be one of $(1, 2), (4, 5)$ or $(9, 10)$.

19: Based on Theorem A of G.H. Hardy, *Notes on some points in the integral calculus*, Messenger of Mathematics 48 (1919), pp.107-112.

20: This identity can be found (eqn. (4.9)) on p.47 of H.W. Gould, *Combinatorial Identities*, Morgantown, W. Va. (1972).

21: The more general equation $a_1x_1 + \cdots + a_nx_n = k$ is treated in Hua Loo Keng, *Introduction to Number Theory*, Springer-Verlag (1982), see Theorem 2.1, p.276.

22: Finite sums of this type are discussed extensively in Chapter 15 of W.L. Ferrar, *Higher Algebra*, Oxford University Press (1950).

25: See Problem 2 on p.113 of W. Sierpinski, *Elementary Theory of Numbers*, Warsaw (1964).

26: Suggested by Problem A-3 of the Forty Seventh Annual William Lowell Putnam Mathematical Competition (December 1986).

29: The discriminant of $f(x^k)$, $k \geq 2$, is given in terms of the discriminant of $f(x)$ in R.L. Goodstein, *The discriminant of a certain polynomial*, Mathematical Gazette 53 (1969), pp.60-61.

30: H. Steinhaus, Zadanie 498, Matematyka 10 (1957), No. 2, p.58 (Polish).

34: This problem was given as Problem 3 in Part B of the Seventh Annual Carleton University Mathematics Competition (1979).

37: Based on a question in the Scholarship and Entrance Examination in Mathematics for Colleges of Oxford University (1975).

38: Based on a question in the Scholarship and Entrance Examination in Mathematics for Colleges of Oxford University (1972).

39: Based on a question in the Scholarship and Entrance Examination in Mathematics for Colleges of Oxford University (1973).

40: Based on a question in the Scholarship and Entrance Examination in Mathematics for Colleges of Oxford University (1973).

41: This is a classical result, see for example H.S.M. Coxeter and S.L. Greitzer, *Geometry Revisited*, Mathematical Association of America (1967), pp.57, 60.

45: Suggested by T.S. Chu, *Angles with rational tangents*, American Mathematical Monthly 57 (1950), pp.407-408.

47: Suggested by W. Gross, P. Hilton, J. Pedersen, K.Y. Yap, *An algorithm for multiplication in modular arithmetic*, Mathematics Magazine 59 (1986), pp.167-170.

48: Based on Satz 3 on p.8 of Th. Skolem, *Diophantische Gleichungen*, Chelsea Publishing Co., New York (1950).

49: Based on Example 1 in D.G. Mead, *Integration*, American Mathematical Monthly 68 (1961), pp.152-156.

52: Suggested by 5.4.5 of L.C. Larson, *Problem-Solving Through Problems*, Springer-Verlag (1983).

53: See Problem 48 of Lewis Carroll's *Pillow Problems*.

56: See Problem 1 on p.13 of W. Sierpinski, *Elementary Theory of Numbers*, Warsaw (1964).

59: See Problem 12 on p.368 of W. Sierpinski, *Elementary Theory of Numbers*, Warsaw (1964).

62: This problem was suggested by Problem A-4 of the Thirty Eighth Annual William Lowell Putnam Mathematical Competition (December 1977).

63: This problem was shown to us by Professors David Richman and Michael Filaseta of the University of South Carolina.

64: This result is due to H.J.S. Smith, *On the value of a certain arithmetical determinant*, Proceedings of the London Mathematical Society 7 (1876), pp.208-212.

68:   This is a well-known problem, see for example 4.4.4 in L.C. Larson, *Problem-Solving Through Problems*, Springer-Verlag (1983).

69:   This problem was suggested by Problem 3 of Part A of the Fifteenth Annual Carleton University Mathematics Competition (1987).

70:   Forms $ax^2+by^2+cz^2$ which represent every integer have been characterized by L.E. Dickson, *The forms $ax^2+by^2+cz^2$ which represent all integers*, Bulletin of the American Mathematical Society, 35 (1929), pp.55-59.

75:   This problem was suggested by Problem 95 of *The Green Book*.

82:   Suggested by K.A. Bush, *On an application of the mean value theorem*, American Mathematical Monthly 62 (1955), pp.557-578.

86:   Suggested by ideas of §7.5, *Estimates of characteristic roots*, in L. Mirsky, *An Introduction to Linear Algebra*, Oxford University Press (1972).

89:   This is a well-known problem. A generalization to the multinomial theorem is given by H.D. Ruderman in Problem 1255, Mathematics Magazine 61 (1988), pp.52-54.

94:   Suggested by an example given in a talk by Professor Michael Filaseta at Carleton University, October 1987.

95:   See Problem 2 on p.219 of W.L. Ferrar, *Higher Algebra*, Oxford University Press (1950).

98:   See Problem 29 on p.123 of E.W. Hobson, *A Treatise on Plane and Advanced Trigonometry*, Dover Publications, Inc. New York (1957).

A CATALOG OF SELECTED
DOVER BOOKS
IN SCIENCE AND MATHEMATICS

A CATALOG OF SELECTED

DOVER BOOKS

IN SCIENCE AND MATHEMATICS

QUALITATIVE THEORY OF DIFFERENTIAL EQUATIONS, V.V. Nemytskii and V.V. Stepanov. Classic graduate-level text by two prominent Soviet mathematicians covers classical differential equations as well as topological dynamics and ergodic theory. Bibliographies. 523pp. 5⅜ × 8½. 65954-2 Pa. $14.95

MATRICES AND LINEAR ALGEBRA, Hans Schneider and George Phillip Barker. Basic textbook covers theory of matrices and its applications to systems of linear equations and related topics such as determinants, eigenvalues and differential equations. Numerous exercises. 432pp. 5⅜ × 8½. 66014-1 Pa. $10.95

QUANTUM THEORY, David Bohm. This advanced undergraduate-level text presents the quantum theory in terms of qualitative and imaginative concepts, followed by specific applications worked out in mathematical detail. Preface. Index. 655pp. 5⅜ × 8½. 65969-0 Pa. $14.95

ATOMIC PHYSICS (8th edition), Max Born. Nobel laureate's lucid treatment of kinetic theory of gases, elementary particles, nuclear atom, wave-corpuscles, atomic structure and spectral lines, much more. Over 40 appendices, bibliography. 495pp. 5⅜ × 8½. 65984-4 Pa. $12.95

ELECTRONIC STRUCTURE AND THE PROPERTIES OF SOLIDS: The Physics of the Chemical Bond, Walter A. Harrison. Innovative text offers basic understanding of the electronic structure of covalent and ionic solids, simple metals, transition metals and their compounds. Problems. 1980 edition. 582pp. 6⅛ × 9¼. 66021-4 Pa. $16.95

BOUNDARY VALUE PROBLEMS OF HEAT CONDUCTION, M. Necati Özisik. Systematic, comprehensive treatment of modern mathematical methods of solving problems in heat conduction and diffusion. Numerous examples and problems. Selected references. Appendices. 505pp. 5⅜ × 8½. 65990-9 Pa. $12.95

A SHORT HISTORY OF CHEMISTRY (3rd edition), J.R. Partington. Classic exposition explores origins of chemistry, alchemy, early medical chemistry, nature of atmosphere, theory of valency, laws and structure of atomic theory, much more. 428pp. 5⅜ × 8½. (Available in U.S. only) 65977-1 Pa. $11.95

A HISTORY OF ASTRONOMY, A. Pannekoek. Well-balanced, carefully reasoned study covers such topics as Ptolemaic theory, work of Copernicus, Kepler, Newton, Eddington's work on stars, much more. Illustrated. References. 521pp. 5⅜ × 8½. 65994-1 Pa. $12.95

PRINCIPLES OF METEOROLOGICAL ANALYSIS, Walter J. Saucier. Highly respected, abundantly illustrated classic reviews atmospheric variables, hydrostatics, static stability, various analyses (scalar, cross-section, isobaric, isentropic, more). For intermediate meteorology students. 454pp. 6⅛ × 9¼. 65979-8 Pa. $14.95

NUMERICAL METHODS FOR SCIENTISTS AND ENGINEERS, Richard Hamming. Classic text stresses frequency approach in coverage of algorithms, polynomial approximation, Fourier approximation, exponential approximation, other topics. Revised and enlarged 2nd edition. 721pp. 5⅜ × 8½.
65241-6 Pa. $15.95

THEORETICAL SOLID STATE PHYSICS, Vol. I: Perfect Lattices in Equilibrium; Vol. II: Non-Equilibrium and Disorder, William Jones and Norman H. March. Monumental reference work covers fundamental theory of equilibrium properties of perfect crystalline solids, non-equilibrium properties, defects and disordered systems. Appendices. Problems. Preface. Diagrams. Index. Bibliography. Total of 1,301pp. 5⅜ × 8½. Two volumes.   Vol. I 65015-4 Pa. $16.95
Vol. II 65016-2 Pa. $14.95

OPTIMIZATION THEORY WITH APPLICATIONS, Donald A. Pierre. Broad-spectrum approach to important topic. Classical theory of minima and maxima, calculus of variations, simplex technique and linear programming, more. Many problems, examples. 640pp. 5⅜ × 8½.   65205-X Pa. $14.95

THE CONTINUUM: A Critical Examination of the Foundation of Analysis, Hermann Weyl. Classic of 20th-century foundational research deals with the conceptual problem posed by the continuum. 156pp. 5⅜ × 8½.   67982-9 Pa. $6.95

ESSAYS ON THE THEORY OF NUMBERS, Richard Dedekind. Two classic essays by great German mathematician: on the theory of irrational numbers; and on transfinite numbers and properties of natural numbers. 115pp. 5⅜ × 8½.
21010-3 Pa. $5.95

THE FUNCTIONS OF MATHEMATICAL PHYSICS, Harry Hochstadt. Comprehensive treatment of orthogonal polynomials, hypergeometric functions, Hill's equation, much more. Bibliography. Index. 322pp. 5⅜ × 8½.   65214-9 Pa. $9.95

NUMBER THEORY AND ITS HISTORY, Oystein Ore. Unusually clear, accessible introduction covers counting, properties of numbers, prime numbers, much more. Bibliography. 380pp. 5⅜ × 8½.   65620-9 Pa. $9.95

THE VARIATIONAL PRINCIPLES OF MECHANICS, Cornelius Lanczos. Graduate level coverage of calculus of variations, equations of motion, relativistic mechanics, more. First inexpensive paperbound edition of classic treatise. Index. Bibliography. 418pp. 5⅜ × 8½.   65067-7 Pa. $12.95

MATHEMATICAL TABLES AND FORMULAS, Robert D. Carmichael and Edwin R. Smith. Logarithms, sines, tangents, trig functions, powers, roots, reciprocals, exponential and hyperbolic functions, formulas and theorems. 269pp. 5⅜ × 8½.   60111-0 Pa. $6.95

THEORETICAL PHYSICS, Georg Joos, with Ira M. Freeman. Classic overview covers essential math, mechanics, electromagnetic theory, thermodynamics, quantum mechanics, nuclear physics, other topics. First paperback edition. xxiii + 885pp. 5⅜ × 8½.   65227-0 Pa. $21.95

TENSOR CALCULUS, J.L. Synge and A. Schild. Widely used introductory text covers spaces and tensors, basic operations in Riemannian space, non-Riemannian spaces, etc. 324pp. 5⅜ × 8¼. 63612-7 Pa. $9.95

A CONCISE HISTORY OF MATHEMATICS, Dirk J. Struik. The best brief history of mathematics. Stresses origins and covers every major figure from ancient Near East to 19th century. 41 illustrations. 195pp. 5⅜ × 8½. 60255-9 Pa. $7.95

A SHORT ACCOUNT OF THE HISTORY OF MATHEMATICS, W.W. Rouse Ball. One of clearest, most authoritative surveys from the Egyptians and Phoenicians through 19th-century figures such as Grassman, Galois, Riemann. Fourth edition. 522pp. 5⅜ × 8½. 20630-0 Pa. $11.95

HISTORY OF MATHEMATICS, David E. Smith. Nontechnical survey from ancient Greece and Orient to late 19th century; evolution of arithmetic, geometry, trigonometry, calculating devices, algebra, the calculus. 362 illustrations. 1,355pp. 5⅜ × 8½. 20429-4, 20430-8 Pa., Two-vol. set $26.90

THE GEOMETRY OF RENÉ DESCARTES, René Descartes. The great work founded analytical geometry. Original French text, Descartes' own diagrams, together with definitive Smith-Latham translation. 244pp. 5⅜ × 8½.
60068-8 Pa. $7.95

THE ORIGINS OF THE INFINITESIMAL CALCULUS, Margaret E. Baron. Only fully detailed and documented account of crucial discipline: origins; development by Galileo, Kepler, Cavalieri; contributions of Newton, Leibniz, more. 304pp. 5⅜ × 8½. (Available in U.S. and Canada only) 65371-4 Pa. $9.95

THE HISTORY OF THE CALCULUS AND ITS CONCEPTUAL DEVELOP-MENT, Carl B. Boyer. Origins in antiquity, medieval contributions, work of Newton, Leibniz, rigorous formulation. Treatment is verbal. 346pp. 5⅜ × 8½.
60509-4 Pa. $9.95

THE THIRTEEN BOOKS OF EUCLID'S ELEMENTS, translated with introduction and commentary by Sir Thomas L. Heath. Definitive edition. Textual and linguistic notes, mathematical analysis. 2,500 years of critical commentary. Not abridged. 1,414pp. 5⅜ × 8½. 60088-2, 60089-0, 60090-4 Pa., Three-vol. set $31.85

GAMES AND DECISIONS: Introduction and Critical Survey, R. Duncan Luce and Howard Raiffa. Superb nontechnical introduction to game theory, primarily applied to social sciences. Utility theory, zero-sum games, n-person games, decision-making, much more. Bibliography. 509pp. 5⅜ × 8½. 65943-7 Pa. $12.95

THE HISTORICAL ROOTS OF ELEMENTARY MATHEMATICS, Lucas N.H. Bunt, Phillip S. Jones, and Jack D. Bedient. Fundamental underpinnings of modern arithmetic, algebra, geometry and number systems derived from ancient civilizations. 320pp. 5⅜ × 8½. 25563-8 Pa. $8.95

CALCULUS REFRESHER FOR TECHNICAL PEOPLE, A. Albert Klaf. Covers important aspects of integral and differential calculus via 756 questions. 566 problems, most answered. 431pp. 5⅜ × 8½. 20370-0 Pa. $8.95

CHALLENGING MATHEMATICAL PROBLEMS WITH ELEMENTARY SOLUTIONS, A.M. Yaglom and I.M. Yaglom. Over 170 challenging problems on probability theory, combinatorial analysis, points and lines, topology, convex polygons, many other topics. Solutions. Total of 445pp. 5⅜ × 8½. Two-vol. set.
Vol. I 65536-9 Pa. $7.95
Vol. II 65537-7 Pa. $7.95

FIFTY CHALLENGING PROBLEMS IN PROBABILITY WITH SOLUTIONS, Frederick Mosteller. Remarkable puzzlers, graded in difficulty, illustrate elementary and advanced aspects of probability. Detailed solutions. 88pp. 5⅜ × 8½.
65355-2 Pa. $4.95

EXPERIMENTS IN TOPOLOGY, Stephen Barr. Classic, lively explanation of one of the byways of mathematics. Klein bottles, Moebius strips, projective planes, map coloring, problem of the Koenigsberg bridges, much more, described with clarity and wit. 43 figures. 210pp. 5⅜ × 8½.
25933-1 Pa. $6.95

RELATIVITY IN ILLUSTRATIONS, Jacob T. Schwartz. Clear nontechnical treatment makes relativity more accessible than ever before. Over 60 drawings illustrate concepts more clearly than text alone. Only high school geometry needed. Bibliography. 128pp. 6⅛ × 9¼.
25965-X Pa. $7.95

AN INTRODUCTION TO ORDINARY DIFFERENTIAL EQUATIONS, Earl A. Coddington. A thorough and systematic first course in elementary differential equations for undergraduates in mathematics and science, with many exercises and problems (with answers). Index. 304pp. 5⅜ × 8½.
65942-9 Pa. $8.95

FOURIER SERIES AND ORTHOGONAL FUNCTIONS, Harry F. Davis. An incisive text combining theory and practical example to introduce Fourier series, orthogonal functions and applications of the Fourier method to boundary-value problems. 570 exercises. Answers and notes. 416pp. 5⅜ × 8½.
65973-9 Pa. $11.95

AN INTRODUCTION TO ALGEBRAIC STRUCTURES, Joseph Landin. Superb self-contained text covers "abstract algebra": sets and numbers, theory of groups, theory of rings, much more. Numerous well-chosen examples, exercises. 247pp. 5⅜ × 8½.
65940-2 Pa. $8.95

Prices subject to change without notice.
Available at your book dealer or write for free Mathematics and Science Catalog to Dept. GI, Dover Publications, Inc., 31 East 2nd St., Mineola, N.Y. 11501. Dover publishes more than 175 books each year on science, elementary and advanced mathematics, biology, music, art, literature, history, social sciences and other areas.